Hyperbolic structures

Matthias Beckh

Hyperbolic structures

Shukhov's lattice towers – forerunners
of modern lightweight construction

WILEY Blackwell Edition **DETAIL**

English language edition first published 2015
© 2015 by John Wiley & Sons, Ltd

Registered Office
John Wiley & Sons, Ltd, The Atrium, Southern Gate, Chichester, West Sussex, PO19 8SQ, United Kingdom

Editorial Offices
9600 Garsington Road, Oxford, OX4 2DQ, United Kingdom
The Atrium, Southern Gate, Chichester, West Sussex, PO19 8SQ, United Kingdom

For details of our global editorial offices, for customer services and for information about how to apply for permission to reuse the copyright material in this book please see our website at www.wiley.com/wiley-blackwell.

The right of the author to be identified as the author of this work has been asserted in accordance with the UK Copyright, Designs and Patents Act 1988.

Originally published in the German language as *Hyperbolische Stabwerke. Šuchovs Gittertürme als Wegweiser in den modernen Leichtbau*
By Matthias Beckh
Published with laminated folded brochure cover in 2012
By Edition DETAIL – Institut für international Architektur-Dokumentation, Munich

This English edition was translated by Raymond D. Peat, Alford, Aberdeenshire (GB).

This book is also available in a German edition (ISBN: 978-3-920034-69-0).

Library of Congress Cataloging-in-Publication Data

Beckh, Matthias, 1974–
  [Hyperbolische Stabwerke. English]
  Hyperbolic structures / Matthias Beckh.
    pages cm
  Includes bibliographical references and index.
  ISBN 978-1-118-93268-1 (pbk. : alk. paper) 1. Hyperbolic structures. 2. Structural frames–Design and construction. 3. Lattice theory. I. Title.
  TA660.H97B4313 2014
  624.1'773–dc23
                          2014026257
A catalogue record for this book is available from the British Library.

Wiley also publishes its books in a variety of electronic formats. Some content that appears in print may not be available in electronic books.

Cover design by Cornelia Hellstern

Set in 8.5/10.7pt Helvetica Neue by SPi Publisher Services, Pondicherry, India

1 2015

Contents

Foreword   8
Introduction   10

Building with hyperbolic lattice structures   14

Geometry and form of hyperbolic lattice structures   24

Structural analysis and calculation methods   32

Relationships between form and structural behaviour   50

Design and analysis of Shukhov's towers   66

NiGRES tower on the Oka   94

Résumé   112

Towers in comparison   114

Notes   143
Literature   145
Picture Credits   146
Index   148

Foreword

The structures of the great Russian engineer Vladimir Grigor'evič Šuchov (Shukhov) are among the world's most sophisticated and distinctive in the history of steel construction. These extremely slender structures, such as cable-stabilised arches, doubly curved gridshells and above all lattice towers hold a great fascination for the observer. They result from the desire to achieve an engineering objective using as little material as possible. At the same time, they are a testament to the extraordinary creativity and inventiveness of an extensively educated engineer, who was on a par with any of his contemporaries.

Many engineering structures of today were anticipated in Shukhov's works. Some of his other structures have no modern equivalent or have remained unmatched in their visual impact and compelling technical efficiency. Among these are without doubt the lattice towers, the foremost being the much photographed and most well-known electricity transmission mast on the Oka (the NiGRES tower) with a height of 130 m and the 150-metre-high Shabolovka radio tower in Moscow, which was originally planned to be 350 m high (Fig. 1). Shukhov built his first hyperbolic lattice tower in 1896 for the All-Russia Exhibition in Nizhny Novgorod. Countless towers with this new type of construction and geometries defined by just a few parameters were built over the following years. These structures proved themselves to be very efficient. The fine-lined tower structures served as water towers, lighthouses, power transmission masts and fire brigade watchtowers – with some of them still in use today.

In his dissertation submitted to the Institute of Structural Design at the Faculty for Architecture, Technical University (TU) Munich, Matthias Beckh analyses these hyperbolic lattice towers for the first time in a systematic manner and investigates the interdependence of form, structure and structural behaviour. The analysis demonstrates how, even in those days, Shukhov was already parametrising his structures as part of the design process. Modern methods of analysis provide not only a gain in scientific knowledge of the lattice structures, they provide Matthias Beckh with the tools to place Shukhov's achievements in a historical context and validate their considerable

contribution to the history of structural engineering. He is also able to demonstrate their relevance to modern structures. All this makes this publication valuable and interesting to a wider public.

Matthias Beckh was part of the first research project into Shukhov's structures, which the Institute of Structural Design undertook together with Rainer and Erika Graefe and Murat Gappoev from the Moscow State University of Civil Engineering. This included investigations of the gridshells for the steelworks in Vyksa and the successful endeavours to make safe and preserve the badly damaged NiGRES tower on the Oka.

Thanks to these, it was possible with the research project into the design knowledge of the early Modern period and Shukhov's strategies for economic steel construction, "Konstruktionswissen der frühen Moderne – V. G. Šuchovs Strategien des sparsamen Eisenbaus", to secure the financial base for an interdisciplinary cooperative project. On this research project, in addition to the Institute for Structural Design, other participants included the Institute for Building History, Building Archaeology and Conservation, TU Munich (Prof. Manfred Schuller), the Institute of Historic Building Research and Conservation, Swiss Federal Institute of Technology (ETH) Zurich (Prof. Uta Hassler) and the Institute for Building History and Conservation, Innsbruck University (Prof. Rainer Graefe). The research project also included in-depth studies of building history as well as dimensional surveys and detailed investigations into the way the structures were built. Later wind tunnel investigations are intended to provide further useful knowledge about the load assumptions, which will also be relevant to modern structures. Finance for the project is provided by the German Research Foundation (DFG), the Swiss National Science Foundation (SNSF) and the Austrian Science Fund (FWF), who are hereby expressly thanked.

Rainer Barthel
June 2012

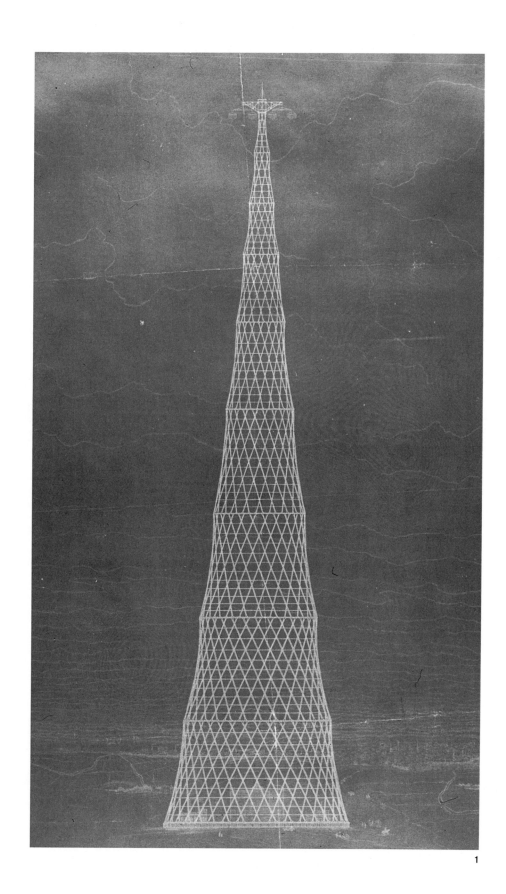

1 Shabolovka radio tower, blueprint of the first (unbuilt) design with a height of 350 m, 1919

1

Introduction

"The Engineer's Aesthetic and Architecture are two things that march together and follow from one another: the one being now at its full height, the other in an unhappy state of retrogression.
The Engineer, inspired by the law of Economy and governed by mathematical calculation, puts us in accord with universal law. He achieves harmony."

Le Corbusier (1887–1965) in *Vers une architecture*, 1923. From a translation of the thirteenth French edition by Frederick Etchtells "Towards a New Architecture".

The object of this book is to gain deeper knowledge on the architectural history of hyperbolic structures. The focus of the investigations is the first hyperbolic lattice towers ever built, the work of Russian engineer Vladimir G. Shukhov. This form of construction, which had no predecessors in the history of building, is notable for its strength and economy of materials. Added to this is the high visual impact of the web-like structures, which almost compel the observer to stand and stare. Even today, Shukhov's load-bearing system can be found in some form or another in architecture, for example in the structural engineering of high-rise buildings.

This book presents the results of the first ever extensive analysis of the way these structures work. The ruled surface of a one-sheeted hyperboloid is resolved into three different mesh variants to create open lattices and their structural behaviour investigated. Then the book looks at the relationships between the load capacities and the basic parameters that determine form as well as the interactions of structural actions and form.

Particular attention is paid to the evaluation and analysis of Shukhov's tower calculations and the assumptions made for the structural model. His historical calculations are compared with the results of modern calculations. Following on from this, Shukhov's design process is reconstructed and the development of the water towers built by him illustrated. The constructional details of the Shukhov-built towers are only touched on here because this subject is currently being more closely examined in an ongoing research project.

Current state of research

Numerous books and papers about Shukhov's work were written in Russia during his lifetime. The papers of Shukhov's biographer Grigorij M. Kovel'man reveal some outstanding insights. Other important contributions that give a good overview of Shukhov's manifold achievements include works by I. J. Konfederatov [1], A. E. Lopatto [2] and Aleksandr Išlinskij [3]. The chapter "Design and analysis of Shukhov's towers" (p. 66ff.) goes into the detail of relevant Russian publications that deal with the structural engineering design and analysis of his towers and the calculation processes used. In the German-speaking parts of the world, Shukhov's works first came to the public's attention in the 1989 book by Rainer Graefe, Ottmar Pertschi and Murat Gappoev "Vladimir Šuchov. 1853–1939. Die Kunst der sparsamen Konstruktion". [4] The wealth of writings by German and Russian authors document the wide variety of his architectural and engineering works.

Until now, the form and geometry of hyperbolic lattice structures have only been investigated to a limited extent. The paper "Zur Formfindung bei Šuchovs mehrstöckigen Gittertürmen aus Hyperboloiden" by Jos Tomlow provides an introduction to the subject. [5] His observations, which were based in part on estimated measurements and geometric parameters, discuss form-finding in the context of structural form, geometry and construction. The relationship between geometry and structural actions is not investigated. Building on this article in his degree thesis, Daniel Günther also discusses the rules of form and geometric dependencies of Shukhovs towers. However, some important relationships are not considered. Although he was the first to evaluate a design table used by Shukhov that gave an insight into his very systematised design process, this analysis did not consider structural behaviour of the towers and its influence on form-finding. [6]

A paper by Peter de Vries discussed the stiffnesses of simple hyperbolic lattice structures and highlighted a single connection between geometry and structural behaviour. The focus here is not on Shukhov, but on a simple form of hyperbolic lattice structure which always has the intermediate rings positioned at intersection points of the lattice members. The results are therefore of secondary relevance to an evaluation of the lattice towers built to the Shukhov design. [7] It can be concluded that, although the geometric relationships of Shukhovs lattice towers have been investigated on various occasions, an analysis of their structural behaviour and the interactions between form and structural behaviour has not taken place to date. However, it is obvious that the form and structure of the Shukhov-built towers were developed not on geometrical or constructional criteria alone, their designs specifically took into account structural engineering considerations. A comprehensive analysis of the way the member arrangements adopted by Shukhov work structurally, of the interactions of form with structural behaviour as well as a reconstruction of the

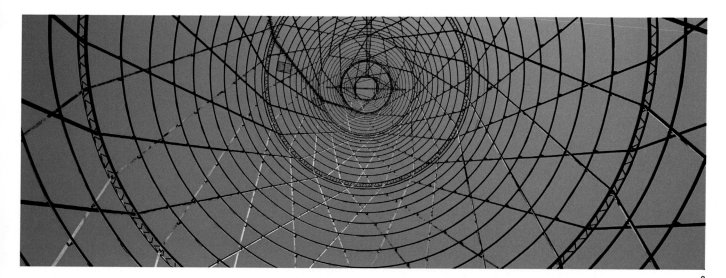

design method derived from these considerations were still lacking. Furthermore there had not been any previous investigations into alternative member arrangements on the surface of a one-sheeted hyperboloid.

Overview

The early chapters of this book set the towers in context with the history of building with iron. This leads on to investigations into the geometric relationships within hyperbolic lattice structures, the way they transfer loads and the interactions between form and structural behaviour. Further key themes include structural calculations and parametric studies, the analysis of some of Shukhov's original structural calculations and his built hyperbolic lattice towers.

The general terms for the structures investigated in this book are "hyperbolic structures" or, slightly more specifically, "hyperbolic lattice structures". When referring specifically to the construction types used by Shukhov, the term "hyperbolic lattice towers" is used. This term is used reasonably often in this context and well established in specialist literature.

In the chapter "Building with hyperbolic lattice structures" (p. 14ff.), the focus falls on the history of building with hyperbolic lattice structures – and the works of the great Russian engineer Vladimir G. Shukhov. After a brief outline of the historic development of building with iron, the discussion centres on Shukhov's diverse contributions in the field of construction and his most important hyperbolic lattice towers.

The chapter "Geometry and form of hyperbolic lattice structures" (p. 24ff.) deals with the form and geometry of hyperbolic lattice structures. A precise description of the mathematical principles of a one-sheeted hyperboloid precedes an explanation of the parametrisation of hyperbolic lattice structures.

Taking these concepts further, the chapter "Structural analysis and calculation methods" (p. 32ff.) considers the principal means of transfer of vertical and horizontal loads and describes the interactions between geometry and structural behaviour. Then follows an explanation of the theoretical principles of determining a lattice tower's ultimate load capacity.

The chapter "Relationships between form and structural behaviour" (p. 50ff.) presents the results of extensive parametric studies of the structural behaviour of hyperbolic lattice structures. Three different arrangements of mesh on the surface of one-sheeted hyperboloids are investigated and the results compared. This is followed by discussions of the calculations for four of Shukhov's built towers.

The chapter "Design and analysis of Shukhov's towers" (p. 66ff.) is devoted to consideration and analysis of Shukhov's structural calculations for the towers. Shukhov's design process is reconstructed based on a comparative evaluation of the historical structural engineering calculations of five different water towers. From the analysis of historic tables stored in the Moscow city archives, a summary of the key data of numerous towers is produced to chart the development of the towers over more than three decades of use.

An analysis of the design and construction of the NiGRES tower on the Oka is the subject of the chapter of the same name (p. 96ff.). The initial ensemble of four electricity transmission masts represents the consummation of Shukhov's tower construction method. Only one of the 130-metre-high masts remains in place today.

A summary of the results of the analysis is provided in the chapter "Résumé" (p. 112f.): It sets out the remaining questions and suggests areas where further research is required.

The final chapter "Towers in comparison" (p. 114ff.) contains an extensive table and drawings of 18 towers.

It is hoped that the examination of Shukhov's form of construction made public in this book will give an impetus to new applications in architecture.

2 Looking up inside the NiGRES tower on the Oka, Dzerzhinsk (RUS) 1929

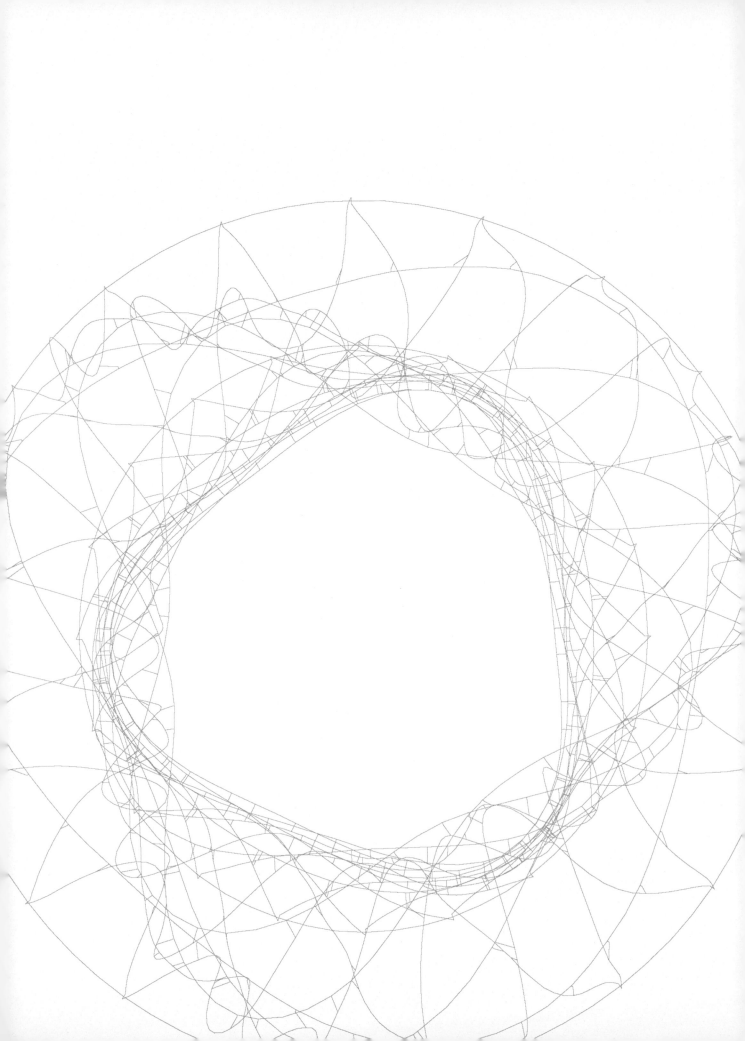

Building with hyperbolic lattice structures   **14**

The development of building with iron in the 19th century   14
The work of Vladimir G. Shukhov, pioneer of lightweight
construction   15
The hyperbolic lattice towers of Vladimir G. Shukhov   19
Hyperbolic structures after Shukhov   23

Geometry and form of hyperbolic lattice structures   **24**

Principles and classification   24
Geometry of hyperbolic lattice structures   28

Structural analysis and calculation methods   **32**

The problem of inextensional bending   32
Principal structural behaviour   32
Theoretical principles for determining ultimate load capacity   38
Parametric studies on differently meshed hyperboloids   45
Principles of the parametric studies   45

Relationships between form and structural behaviour   **50**

Comparison of circular cylindrical shells and hyperboloids
of rotation   50
Mesh variant 1: Intermediate rings at intersection points   50
Mesh variant 2: Construction used by Vladimir G. Shukhov   52
Mesh variant 3: Discretisation of reticulated shells   57
Summary and comparison of the results   59
Structural analysis of selected towers built by Vladimir
G. Shukhov   60

Design and analysis of Shukhov's towers   **66**

The development of steel water tanks and water towers   66
The water towers of Vladimir G. Shukhov   69
Development of structural analysis and engineering design
methods in the 19th century   70
Calculations for Vladimir G. Shukhov's lattice towers   70
Evaluation of the historical calculations   84
The design process adopted by Vladimir G. Shukhov   89

Building with hyperbolic lattice structures

Building with hyperbolic lattice structures began with the Russian engineer and polymath Vladimir Grigor'evič Shukhov (1835–1939). After a short summary of the earlier development of building with iron, this chapter will cover the history of this form of construction.

The development of building with iron in the 19th century

The threshold of the 19th century saw many new types of different load-bearing structures arising in western Europe. The conditions for this turn of events were created from the end of the 18th century by the rapid pace of industrialisation and the material being used to make new tools and machines – iron. New methods of production of this metal meant that ample quantities were also available for construction. After thousands of years of the predominance of stone and wood in buildings, architects and engineers were able to use iron not just as a means of making connections but also as a construction material in its own right. The characteristic properties of the new construction material, in particular its high strength, the toughness of wrought iron and, in contrast to the latter, the brittleness of cast iron demanded new types of construction and details. At the same time, the beginning of the 19th century presented architects and engineers with construction tasks quite unheard of before. Spacious railway stations, large exhibition halls for industrial expos and glazed arcades called for new structural solutions to bridge the often considerable spans. The new material's high strength gave designers the opportunity to realise relatively lightweight and delicately proportioned structures.

But designing in iron was destined to have a long developmental phase. Until the middle of the 19th century, engineers cautiously felt their way forward with new types of structures and methods of construction, which led the first iron roof trusses to look very much like their wooden predecessors. However, the creation of iron structures required a complete rethinking of design and construction planning: The necessary prefabrication of the structural members and their details in factories was shifting the focus of the building process from the construction site to the workshop; factory assembly was replacing the flow of conventional skilled craftsmen's operations on site. The expense and effort required to make casting moulds and fabricate connections forced designers to repeat elements as many times as possible. Early consideration of how members could be joined to one another efficiently and erected quickly was increasingly influential in the design and led to new forms of construction, systems and details: "Repetitive elements and standardized connections characterize a system approach to design that implies organizing component hierarchies rather than composing forms," writes Tom F. Peters in "Building the Nineteenth Century". [1] Modular building systems like the one used for the Crystal Palace, built to house the Great Exhibition of 1851, in London epitomise this development (Fig. 1). The details in these structures became more and more sophisticated, not only successfully contributing to the continuity of form and load transfer but also fulfilling a wide range of other requirements, such as the ability to accommodate temperature fluctuations and fit in with the sequence of operations on site. [2]

Among the many progressive building projects completed by the middle of the century were the cupola of the corn exchange (Halle au blé) by François-Joseph Bélanger and François Brunet in Paris (1811), the casting shop at the Sayn ironworks in Bendorf by Carl Ludwig Althans (1830) and the Palm House in Kew Gardens in London by Richard Turner (1848), all of which have primary load-bearing elements made from cast iron. Developments in bridge-building also had an impact on buildings, for example cast-iron arches, which are usually composed of several segments. New systems appeared, such as the Wiegmann-Polonceau girder, which can be seen in countless railway stations and market halls, and its further development, the sickle girder, which was first used by Richard Turner in Lime Street Station in Liverpool (1849). [3]

"The 1840s marked the end for the first epoch of iron construction, which had been very largely cast iron based," remarks Werner Lorenz in his book "Konstruktion als Kunstwerk". [4] Following the invention of the Bessemer (1856) and the Siemens-Martin smelting processes

Hyperbolic structures: Shukhov's lattice towers – forerunners of modern lightweight construction, First Edition. Matthias Beckh.
© 2015 John Wiley & Sons, Ltd. Published 2015 by John Wiley & Sons, Ltd.

(1864), wrought iron and eventually steel became cheap and available in sufficient quantities. From 1845, it was possible to produce rolled I-sections, which quickly became very popular. In addition to these technical and industrial advances, the use of the new materials was accelerated further thanks to the development of structural mechanics, which was increasingly seen as a separate engineering discipline. Largely responsible for the recognition of this new discipline was Claude Louis Navier, through his outstanding paper on the elastic behaviours of structures (originally published in French in 1826 [5]), which was available in most European languages from the middle of the century, alongside publications on the theory of trussed frameworks by Johann Wilhelm Schwedler and Carl Culmann (1851) [6], with Culmann also being responsible for developing the technique of graphic statics (1862). These technical and scientific innovations encouraged the development between 1850 and 1880 of many new load-bearing systems whose dimensions were no longer determined empirically or intuitively as before but by calculation. At the heart of the development were the countless new truss systems that came to the fore at this time. "If the fifties were the decade of cautious exploration of new systems, then the sixties began the classic era of trusses in buildings." [7] The railway station halls of this time in metropolises, such as London (e.g. St. Pancras and Victoria stations) and Paris, bear clear witness to the advances in spanning capability and efficient use of materials. Further new load-bearing systems that became very popular because they are statically determinate and therefore simple to design included suspended span girders (also known as Gerber beams) and three-pinned arches, which were used in buildings for the first time by Schwedler in 1865 and set against a splendid backdrop in Charles Louis Ferdinand Dutert and Victor Contamin's Galerie des Machines (Fig. 2), built for the Paris International Exhibition in 1889 [8]. In 1863, Schwedler's success with his "Schwedler domes" was the breakthrough for three-dimensional reticulated shells.

In the last quarter of the 19th century, the pace of innovation in iron and steel construction in western Europe hit a plateau. While new records for girder spans and building heights continued to be set, the development of a canon of new types of structures was more or less

at an end, their design mastered. The eyes of engineers in western Europe were now focused mainly on a new material developing at a tremendous pace at this time: reinforced concrete.

The important stimulus to architecture and its revival that the new style of engineering structures of the 19th century and their aesthetics gave is underlined by a quotation from Henry van de Velde, the Belgian architect and designer who was involved in the Bauhaus movement. Writing on the role of the engineer in 1899, he said "there is a class of people from whom the title artist can no longer be withheld. These artists, these creators of the new architecture, are the engineers. The extraordinary beauty inherent to these works of engineers, is based on an unawareness of their artistic possibilities – as it is with the creators of the beauty of our cathedrals, who were also unaware of the magnificence of their works." [9]

The work of Vladimir G. Shukhov, pioneer of lightweight construction

The last quarter of the 19th century was a period of advancing influence for the Russian engineer and inventor Vladimir G. Shukhov, one of the most important pioneers of lightweight construction and modern building with iron and steel. Shukhov was as significant to the development of lightweight structures as outstanding engineers such as Robert Maillart or Pier Luigi Nervi were to the advancement of modern reinforced-concrete construction. However, it is Shukhov's extraordinary versatility which allows him to stand comparison with similar universally proficient engineers at the end of the 19th century such as Alexander Graham Bell or Gustave Eiffel.

A wealth of design principles, which are still applied in structural steelwork today, find their roots in Shukhov's works. He was responsible for the first doubly curved gridshell, built the first suspended roofs and developed extremely slender arched girders, which were

1   Crystal Palace, London (GB) 1851, Joseph Paxton, interior view from "The Crystal Palace Exhibition Illustrated Catalogue", London 1851
2   Galerie de Machines, Paris (F) 1889, Charles Louis Ferdinand Dutert, Victor Contamin

3 Schematic section (a) and interior photograph (b) of the arcade roofs of the GUM department store, Moscow (RUS) design from 1890
4 Gridshell mesh roof over a pump station, Grozny (RUS) ca. 1890
5 3D visualisation of the doubly curved gridshell, Vyksa (RUS) 1897

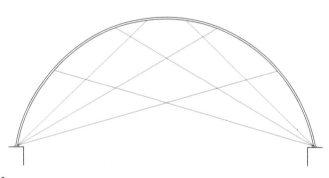

a

b   3

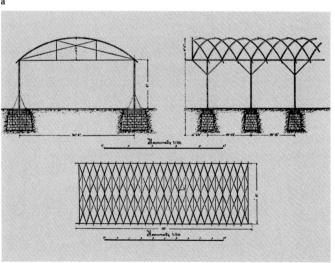

4

5

stiffened by thin tensile members. But it is not this technical finesse alone which makes his structures so fascinating. The delicate, almost dematerialised, structures have a high degree of aesthetic attraction from which it is difficult to disengage. In his homeland, Shukhov was decorated with the highest national awards and is still spoken of today in modern Russian society. However, outside Russia his diverse and extraordinary works remain for the most part unknown. It was not until the publication in 1989 of "Vladimir G. Šuchov. 1853–1939 Die Kunst der sparsamen Konstruktion" [10] by Rainer Graefe, Murat Gappoev and Ottmar Pertschi that the interest of the broader professional public was awoken in western Europe.

Outline biography

Shukhov was born in 1853 in the small town of Grajvoron near the Ukrainian border and his youth coincided with a time of great social and economic upheaval in Russia. Alexander II implemented numerous reforms in the backward tsardom. Among other measures, he abolished serfdom in 1861 and reformed the universities. The monopolistic, state-controlled policies produced an upturn in the mining and heavy industries and a gain in the pace of industrialisation in Russia, which had been rather late in starting. In this time of rapid change, the 18-year-old Shukhov enrolled at the polytechnic in Moscow, which had emerged from the earlier state craftsmanship schools and offered a very advanced curriculum. In addition to a thorough study of mathematics and physics, his education also gave him a practical grounding in the polytechnic's own workshops. In 1878, two years after his graduation as a "Mechanic Engineer", Shukhov began his professional life at the engineering company Alexander V. Bari in Moscow, which he had come across at the International Exhibition in Philadelphia and would be associated with for much of his life. Soon after his appointment, Shukhov rose to become chief engineer and was named company manager in 1918 following nationalisation. [11]

His first assignment with Bari took him to the Russian colony of Azerbaijan, where he quickly made a crucial contribution to the expansion of the Russian oil industry: he developed an industrial plant for the thermal cracking of crude oil, designed the first Russian pipeline and constructed the world's first pipeline for preheated mazut, a viscous liquid residue from petroleum distillation. At the same time, he designed and built the first cylindrical crude oil tanks, the same principle being still in use today. On his return to Moscow, he continued to work with the same intensity. In 1879, Shukhov patented a nozzle for burning mazut, and in 1885 he designed the first Russian tanker ships. A few years later, he filed patents for horizontal and vertical boilers for which he would receive numerous awards. The abundance of his engineering innovations and inventions seems almost limitless, as do his fields of activity. From 1890, Shukhov turned his attention increasingly towards buildings. The following sections cover his most pioneering innovations in the field of construction engineering. [12]

Arched girders

In 1890, Shukhov started work on the design of the barrel-shaped roofs for arcades in the GUM department store, situated directly adjacent to Red Square in the centre of Moscow (Fig. 3). Their semicircular arched elements bridge the 15 m spans and have a

sophisticated stiffening system: Shukhov attached three slender ties to each impost, which were then fanned out and connected to the arch.

To understand the way this completely new method of construction worked requires some familiarity with the deformed shape of an arch. Arched girders are particularly efficient when supporting uniformly distributed loads, but asymmetric loads, such as wind or a one-sided snow load, cause the arch to deform out of line; only the flexural stiffness of the arch then resists the deformation. This necessarily higher stiffness can only be achieved by having a larger cross section, which makes the structure relatively heavy.

The radial arrangement of ties is designed specifically to prevent this deformation. As a result, the cross sections can be kept small. Shukhov designed almost all his arched girders according to this principle. After this method of construction had been long forgotten, several structural engineers in the 1990s rediscovered and further developed his design approach. Examples include the station roof in Chur by Peter Rice and the radially stiffened arch elements frequently used by Jörg Schlaich for his reticulated shells.

Lattice gridshells

1890 was also the year Shukhov built the first lattice gridshells. While the reticulated shells of Berlin engineer Johann Wilhelm Schwedler, which were known as "Schwedler domes", had members of different sizes to suit their loads, Shukhov's lattice shells were made up of elements of the same size. For the roof of a pump station in Grozny, he used two layers of circular arc segments made from steel profiles that ran from edge beam to edge beam to create a shape in plan composed of rhomboid meshes (Fig. 4). The beams, fabricated out of Z-profiles, are riveted together at the intersection points, the shear forces are carried by horizontal circular rods. The radial stiffening system of the arched girder mentioned above is used again here to stabilise the singly curved surface. [13]

The contract for the temporary exhibition hall for the All-Russia Exhibition in Nizhny Novgorod helped this form of construction to make its final breakthrough. Bari constructed four large halls with more than 16 000 m² of exhibition space using gridshells. The popularly named "roofs without trusses" [14] achieved spans of up to 32 m. Bari's drawings demonstrate that a great many warehouses and factory buildings with a variety of spans, rises and numbers of ties were built in Russia using this basic design.

The hall designed by Shukhov in 1897 in the metal production city of Vyksa reached a whole new level of quality. The construction principle of the barrel-shaped gridshell roof developed for Grozny is led along a parabolic – the first doubly curved gridshell in the world. Analysis of the structure and the historic calculations shows that the double curvature of the gridshell was chosen for practical construction reasons and not on structural engineering grounds. [15] Today, after lying unused for 20 years, the much neglected hall is in urgent need of refurbishment. It has been extensively investigated and recorded in recent years. [16] Plans for refurbishment and new use concepts are already being discussed. The unusually light construction has lost none of its elegance, even 110 years since its erection, as the visualisation of the internal space shows (Fig. 5, p. 16). [17]

6 Doubly curved glass roof of the British Museum, an example of a modern reticulated shell, London (GB) 2000, Foster and Partners
7 Cross section through the suspended roof on the rotunda at the All-Russia Exhibition in Nizhny Novgorod (RUS) 1896
8 Drawing from Shukhov's patent application No. 1896

6

Suspended roofs

In 1895, Shukhov registered a patent for a suspended mesh roof in which he transformed his compression-loaded gridshell system into a tensile structure: he suspended two layers of steel flats, skewed in opposite directions to one another, between a tensile top and a bottom compression ring. In plan, the meshes were the characteristic diamond shape. He created the world's first steel anticlastically curved tensile structure. Just as he did with the gridshells, Shukhov used this new form of construction to span more than 10000 m² at the All-Russia Exhibition, this time with four suspended roofs. The most spectacular example is the rotunda with a diameter of 68 m, which is a combination of two suspended roofs (Fig. 7). Between a higher ring supported by 16 lattice columns and a compression ring on the outside are two families of steel flats, which create a fine mesh and span the 21.5 m between the two rings. The dimensions of the 640 steel strip sections are only 50 ≈ 5 mm. The interior of the rotunda is also covered by a tensile structure. Though here, the load-bearing structure is not a mesh, but a thin membrane. A suspended calotte of riveted sheets, a mere 1.6 mm thick, hangs over the 25 m diameter inner circle. This building therefore used two tensile structures whose horizontal support forces to some extent cancelled one another out. [18]

In spite of their low self-weight, no more suspended grid roofs were built in the years after the All-Russia Exhibition. This form of construction seemed to have been forgotten until almost 60 years later when the Polish-American architect Matthew Nowicki designed the famous suspended cable net for the Dorton Arena in Raleigh in 1953, which prepared the way for Frei Otto and his cable net structures.

The hyperbolic lattice towers of Vladimir G. Shukhov

In 1896, Shukhov developed a new type of structure which had no predecessors in building history: the hyperbolic lattice tower. The doubly curved surface of a one-sheeted hyperboloid can be created by rotating a skewed line about a vertical axis. Shukhov put this principle to use and created a mesh of two counter-running families of straight members on this surface.

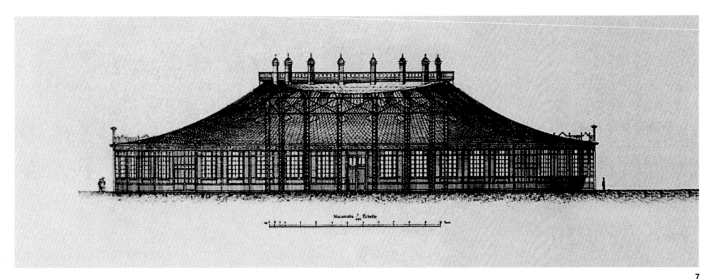

The resulting structure is relatively light and exceedingly efficient; it is also simple and quick to erect. For these reasons, but also for its characteristic shape, it was used extensively in Russia and her colonies – mainly for the countless water towers built according to this principle. The method of construction was also used for shot towers (used for the production of lead shot) and lighthouses as well as radio and electricity transmission masts. The structures were also used as lookout masts on ships of the Russian and United States navies. [19]

Patent No. 1896

On 11th January 1896, Shukhov submitted a patent application in Moscow for his new type of tower structure, which was granted on 12th March 1899 (Fig. 8). He described the content of the patent in the following words: "A lattice-form tower characterised in that its load-bearing structure consists of straight wooden beams, iron tubes or angle profiles which cross over one another and lie on the directrix of a solid of revolution and that takes the form of a tower. They are riveted to one another at the crossing points and also connected by horizontal rings."

Shukhov described the advantages of this method of construction thus: "The tower built in this way is a stable structure which resists extreme forces and uses very little material. The main application of these structures could be as water towers or lighthouses." [21]

How Shukhov arrived at his invention cannot be precisely reconstructed from the available records. Grigorij M. Kovel'man wrote on the subject: "I thought a long time about the hyperboloid. Then something evidently took place in my subconscious but did not spring directly from it." [22] As well as the theory that a small wicker basket in the form of a hyperboloid, which Shukhov used in his office as a waste paper bin, could have served as the inspiration, it may also have been simply the sound teaching during his studies that provided the basis for the idea. Shukhov recalled "in the lectures on analytical geometry, it was said that hyperboloids were good training for the intellect, but had no practical uses." [23] Another supposition is that an important source of inspiration for

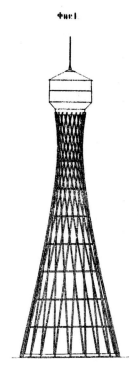

Фиг. 1.

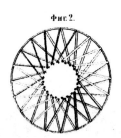

Фиг. 2.

8

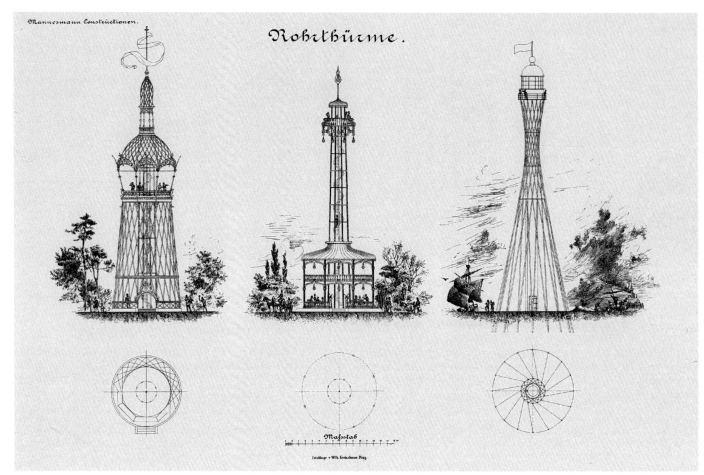

Mannesmann Constructionen.

Rohrthürme.

Maßstab

9   Mannesmann tube towers, undated
10   First hyperbolic lattice tower by Shukhov at the exhibition in Nizhny Novgorod
     (RUS) 1896 (a) and at its present location in Polibino, southern Russia (b)
11   Water towers in Kolomna (RUS) 1902 (a), Mykolaiv (UA) 1907 (b), Kharkiv
     (UA) 1912 (c) and Džebel (TM) 1912 (d)

Building with hyperbolic lattice structures

Shukhov was the non-Euclidian geometry developed by the Russian mathematician Nikolai Lobatschewski in 1829. [24] This theory, which has no further substantiation, would be very difficult to verify in any case.

The most probable assumption is that he had the idea of the hyperbolic tower after extensive consideration of the suspended roofs he developed earlier: if the curved steel strips of his rotunda roof are replaced by straight members, it results in the new form of structure.

The tubular towers of Mannesmann

In 1885, brothers Reinhard and Max Mannesmann invented the process of skew rolling to manufacture seamless pipes. This very quickly transformed the small Remscheid file-making factory into a major company with subsidiaries and business dealings all around the world. In the unpublished documents of Reinhard Mannesmann, which can be found in the archives of the German Museum in Munich, there is a drawing of lookout towers made from tubular profiles, two of which are in the form of a hyperboloid (Fig. 9). According to the information in the archive, the drawing, which is not dated, must have been prepared between 1890 and 1895 and had already been published in the papers of Ruthild Brandt-Mannesmann and Berthold Burkhardt. [25] The extent to which the drawing had been been in circulation at the time can no longer be determined.

Considering the enormous importance seamless tubes would have been in particular to machine making – in comparison with the welded alternatives, they are more reliable and can withstand higher pressures – and the fact that Mannesmann supplied the pipes for the first pipeline from Balachna to Cherny Gorod – with Bari as the main contractor for Nobel Brothers – there would seem to be a great deal of overlap between Shukhov's spheres of activity. [26] Possibly Shukhov knew of the designs for the Mannesmann tubular towers, but there is no evidence for this. The Mannesmann tubular towers themselves were never built; the preliminary design was more likely a demonstration of the diverse fields of application of the new product. What cannot be disputed in any case is that Shukhov was the first to build hyperbolic lattice towers.

Water towers

Soon after the submission of patent application No. 1896, Shukhov built his first tower using his new principle. He erected a water tower with a height of 25.6 m and a shaft with 80 straight members made out of angle profiles for the All-Russia Exhibition in Nizhny Novgorod (Fig. 10a, p. 21). The tower, which was used by Bari as a showpiece, caused great excitement among the visitors and in the international specialist press. A particular attraction was the Moiré effect, which appears to make the structure move. [27] After the exhibition, the tower was translocated to a new site in Polibino in southern Russia, where it stands today (Fig. 10b, p. 21).

The construction of this prototype signalled the start of a real boom in the following years. According to some sources, over more than 200 towers of this new design were erected in all parts of Russia;

a

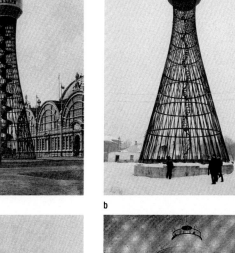

b   10

a

b

c

d   11

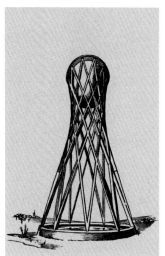

12   13   14   15

the precise number can no longer be identified. There are two crucial reasons for this: firstly the economically efficient construction of the towers, which very often considerably undercut the competing systems, as was demonstrated by the water tower in Mykolaiv, for which Bari tendered and won the contract. Secondly the high architectural quality of the towers was crucial for this success: "The search for new engineering solutions was fuelled by prosperous cities in the 19th century using these utility structures as architectural attractions. In the mainly one- or two-storey Russian cities, these high water towers, masts and lighthouses must have provided an element urban beauty." [28] The water tower built in a prominent position near the church in Kolomna (Fig. 11 a, p. 21) underlines this statement. Perhaps the historic photograph also shows a somewhat less prejudiced and more open attitude existed in Russia towards the new types of engineering structures than that prevalent at that time in Western countries, where the erection of a modern utility building in the historic centre of a historically significant city would certainly have met with resistance.

In spite of the large number of towers built and the highly systemised design process, hardly any two towers are exactly the same. Geometry and structural members are determined specifically for each project. The chapters "Relationships between form and structural behaviour" (p. 50ff.) and "Design and analysis of Shukhov's towers" (p. 66ff.) deal exclusively with the construction, structural analysis and the design of water towers and their development.

Multistorey towers

As well as single-storey towers, there were also towers constructed from more than one hyperboloid segment placed one on top of the other. The multistorey construction technique was used for the first time on the water tower erected in 1911 in Yaroslavl. A suspended-bottom tank is installed at the top of each of the two segments, which each consist of 30 vertical members. The tower has a total height of 39.1 m at the top ring. A handwritten fragment of the calculations for this tower is discussed in the chapter "Structural calculations for the two-storey water tower in Yaroslavl" (p. 79f.).

Shabolovka radio tower

In 1919, shortly after the revolution, the demand for lattice towers took on another order of magnitude: Shukhov designed a radio transmission tower for the Comintern radio station Shabolovka in the centre of Moscow, which was to provide the radio broadcasting link from the capitol to the provinces. The first design was for a 350-m-high tower made from nine hyperboloid segments (Fig. 1, p. 9). In contrast to the other multistorey towers, the number of vertical members reduces continuously from 72 in the first segment to 12 in the ninth. [29] The even decrease in the number of vertical members creates eccentricities at the transitions between the individual sections. This structural problem had to be solved by installing spreaders under each main ring to bridge across the ends of the vertical members. A shortage of structural steel prevented this bold design from being built to its original ambitious dimensions. Eventually a 150-metre-high version was completed in 1922 (Fig. 12). It has 48 vertical members in the lower four segments, halving to 24 in the upper two. The type and size of the steel profiles in each segment are designed to suit the actual load. The structure was erected using the telescopic method in which the next hyperboloid segment is assembled at ground level inside the shaft and then hoisted upwards using small timber industry cranes to rest on top of the last section.

The completion of the structure, at that time the tallest in Russia, was celebrated as the "Trumpet of the radio revolution" [30] and was mentioned in contemporary works of literature and the performing arts. The tower remained the tallest structure in Moscow for many decades. Even today, its signature silhouette characterises this city.

NiGRES towers on the Oka

Perhaps the most beautiful of Shukhov's tower structures are the NiGRES towers erected as an ensemble of pairs of high-voltage electricity transmission masts in 1927 southwest of Nizhny Novgorod on the Oka. The mast pairs were stepped in height and reached 130 m at the river bank to allow the conductors to span almost 1000 m across the water. Nothing before had matched this late

16

work of Shukhov for its timeless structural elegance and the natural simplicity of its detailing. The design and load-bearing structure of this tower are discussed extensively in the chapter "NiGRES tower on the Oka" (p. 96ff.).

Hyperbolic structures after Shukhov

Hyperbolic lattice structures were very seldom built after Shukhov's death in 1939. A few structures have been designed or built in the intervening 70 years that made reference – at least in terms of form – to Shukhov's invention. The most famous are the water tower feasibility study by the Spanish engineer Eduardo Torroja (Fig. 13) and the unbuilt skyscraper for mid-town Manhattan designed in 1954 by Ieoh Ming Pei (Fig. 14). In addition, several high-rises have emerged with load-bearing systems based on hyperbolic lattice structures. Modern examples in Ghuangzhou – the highest television tower in the world at 610 m (Fig. 16) – and in Doha (Fig. 17) show the endless fascination of these structures.

12 Shabolovka radio tower, Moscow (RUS) 1922
13 Preliminary design of a water tower, 1935, Eduardo Torroja
14 Unbuilt high-rise, 1954, I. M. Pei
15 Mae West sculpture, Munich (D) 2011, Rita McBride
16 Ghuangzhou TV Tower (CN) 2010, IBA Information Based Architecture
17 Tornado Tower, Doha (Q) 2008, SIAT – Architeckten und Ingenieure; CICO Consulting Architects Engineers

17

Geometry and form of hyperbolic lattice structures

1

Although the form of a hyperbolic lattice structures appears complex at first sight, its geometry and the positions of its members can be unambiguously defined using only a few basic parameters. This chapter illustrates the geometrical rules.

Principles and classification

One-sheeted hyperboloids are second-order algebraic surfaces, also known as quadratic surfaces. This chapter summarises their most important principles and geometric relationships as far as they are relevant to hyperbolic structures.

Conic sections

All plane, quadratic curves (parabola, hyperbola and ellipse) can be represented with the help of conic sections. Conic sections are the curves obtained from the intersection of a plane with a cone (Fig. 1). The relationship between the half-aperture angle of the cone α and the angle between the intersecting plane and the cone axis β can be defined with the following formula:

$$\nu = \frac{\cos \beta}{\cos \alpha} \qquad \text{(F 01)}$$

If $\nu > 1$, then the created intersection curve is a hyperbola, if $\nu = 1$, the curve is a parabola and if $\nu < 1$ it is an ellipse. The Greek names of conic sections are based on the relationship of ν to the number 1: With the hyperbola (from the Greek for "excess") the value of the quotient ν exceeds the number 1, with the ellipse (from the Greek for "falls short") it is less than 1, while with the parabola (from the Greek for "a placing alongside") it equals 1. The discovery of conic sections is attributed to the Greek mathematician Menaechemus, a member of the Platonic Academy; their most important characteristics were defined by the Greek mathematician Apollonios of Perga in his book "Conics" [1].

Second-order algebraic surfaces

A second-order algebraic surface or quadric is described as a set of all the points $X = (x_1, x_2, \ldots, x_n)$ that satisfy an equation of the following conventionally written form:

$$p(x) = x^T A x + a^T x + \beta = 0,$$
$$\text{with } X = (x_1, \ldots, x_n) \in R_n, \qquad \text{(F 02)}$$
$$A = A^T \in R^{n\times n}, a \in R^n, \beta \in R$$

From a total of 17 second-order algebraic surfaces, five are doubly curved, four are singly curved, the rest are planes, straight lines or points or are empty sets. The doubly curved second-order algebraic curves can be classified into two groups according to how they are generated: doubly curved revolution surfaces (Fig. 2) and doubly curved translation surfaces (Fig. 3). Translating a hyperbola along a second creates a two-sheeted hyperboloid; the translation of a parabola along a second creates an elliptic paraboloid. If the direction of one of this pair of parabolas is reversed, the resulting form is a hyperbolic paraboloid. Rotation of a hyperbola likewise creates a one- or two-sheeted hyperboloid. Rotating a parabola about its axis of symmetry creates the surface of a rotation paraboloid. On the other hand, rotating of an ellipse about one of its two axes of symmetry creates a rotation ellipsoid. [2]

One-sheeted hyperboloid

There are three different ways of unambiguously describing the surface of a one-sheeted hyperboloid (Fig. 5, p. 26). The first is by rotating a skew line about an axis, the second is the rotation of a hyperbola about an axis and the third is to create three lines skewed to one another, which always uniquely define a hyperboloid of one sheet.

In a similar form to equation F 02, the one-sheeted hyperboloid (Fig. 4 b) can be defined in conventional terms [3] with the following equation:

$$\frac{x^2}{a^2} + \frac{y^2}{b^2} - \frac{z^2}{c^2} = 1 \qquad \text{(F 03)}$$

The first two terms correspond with the equation for an ellipse and describe the form of the hyperboloid in plan. For a hyperboloid that is circular in plan, b equals a, which considerably simplifies the equation. The variables a and b therefore define the semi-major axes of the "waist" in the x and y directions. The variable c defines

Hyperbolic structures: Shukhov's lattice towers – forerunners of modern lightweight construction, First Edition. Matthias Beckh.

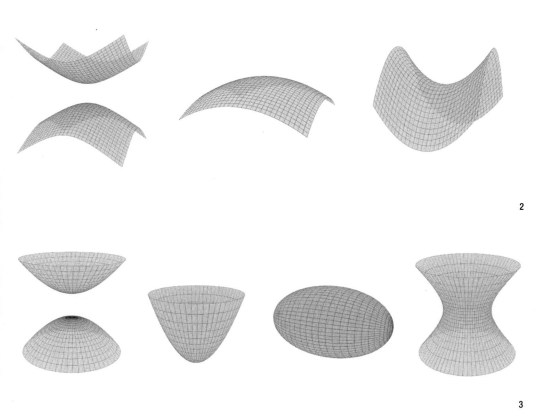

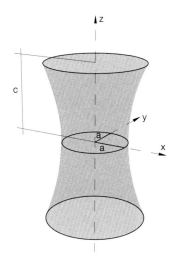

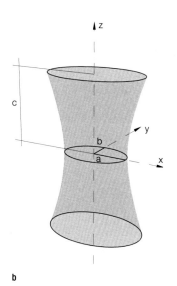

1 Conic sections (l. to r.): hyperbola,
 parabola and ellipse
2 Doubly curved second-order
 rotation surfaces: two-sheeted
 hyperboloid, rotational paraboloid,
 spheroid and one-sheeted hyperboloid
3 Doubly curved second-order translation surfaces: two-sheeted hyperboloid, elliptical paraboloid and
 hyperbolic paraboloid
4 Axes of rotational hyperboloid (a)
 and general one-sheeted hyperboloid (b)

a

b

4

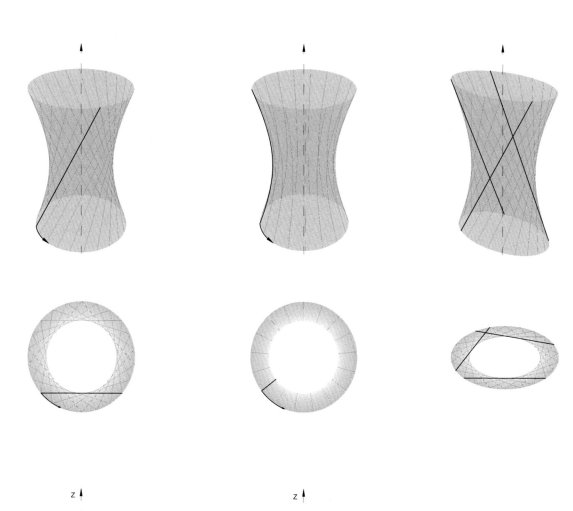

5

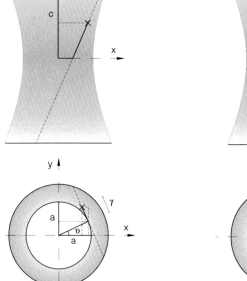

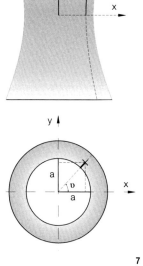

6

7

5   One-sheeted hyperboloid: different
    methods of generation
6   Generation of a one-sheeted
    hyperboloid by rotating a straight
    line generatrix
7   Generation of a one-sheeted
    hyperboloid by rotating a
    hyperbola

Geometry and form of hyperbolic lattice structures

the necking in the elevation of the hyperboloid. The greater the value of c, the smaller the amount of necking.

In addition to the standard and various different possible parameter notations, the surface of the general one-sheeted hyperboloid can also be described using the following general cylindrical coordinate system:

$$
\begin{aligned}
x &= a\sqrt{1+u^2}\cos\upsilon \\
y &= b\sqrt{1+u^2}\sin\upsilon \\
z &= cu \\
\upsilon &\in [0, 2\pi], u \in N
\end{aligned}
\qquad \text{(F04)}
$$

In the case of a rotation hyperboloid (Fig. 4a, p. 23), both the semi-major axes are the same size, which therefore allows the standard notation to be simplified to:

$$
\frac{x^2}{a^2} + \frac{y^2}{a^2} - \frac{z^2}{c^2} = 1
\qquad \text{(F05)}
$$

The radius of any section at height z is therefore:

$$
R = a\sqrt{1 + \frac{z^2}{c^2}}
\qquad \text{(F06)}
$$

Creation by rotation of a skewed straight line

A one-sheeted hyperboloid can be created by rotating a straight line generatrix about a straight line at a skew to the rotation axis (Fig. 6). Rewriting equation F 05 gives two general linear equations for the two families of straight lines rotating in opposite senses. [4]

Using the free parameters u and v, the equations take the form in standard notation:

$$
\frac{x}{a} + \frac{z}{c} = u\left(1 + \frac{y}{b}\right), \quad u\left(\frac{x}{a} - \frac{z}{c}\right) = 1 - \frac{y}{b};
\qquad \text{(F07)}
$$

$$
\frac{x}{a} + \frac{z}{c} = v\left(1 - \frac{y}{b}\right), \quad v\left(\frac{x}{a} - \frac{z}{c}\right) = 1 + \frac{y}{b};
\qquad \text{(F08)}
$$

A parametric representation with which every point on the surface of the hyperboloid can be described follows the paths of the generating straight lines. The position of any point can be determined from the angle υ at which the relevant generatrix lies at a tangent to the waist circle and the horizontal projection γ of this generatrix. As there are two families of generating straight lines, each point is controlled by two straight lines. The coordinates of the points can be calculated from:

$$
\begin{aligned}
x &= a\,(\cos\upsilon \pm \gamma\sin\upsilon) \\
y &= a\,(\sin\upsilon \pm \gamma\cos\upsilon) \\
z &= c\gamma
\end{aligned}
\qquad \text{(F09)}
$$

The semi-major axis c is therefore synonymous with the slope of the straight line generatrices.

Creation by rotating a hyperbola

A one-sheeted hyperboloid can also be created by rotating a hyperbola with the equation:

$$
\frac{x^2}{a^2} - \frac{z^2}{c^2} = 1
\qquad \text{(F10)}
$$

about the vertical z-axis (Fig. 7). This allows the derivation of the parametric equation [5] based on the inverse sine and cosine hyperbolic functions:

$$
\begin{aligned}
x(u, v) &= a\cos\upsilon\cosh\xi \\
y(u, v) &= a\sin\upsilon\cosh\xi \\
z(u, v) &= c\sinh\xi
\end{aligned}
\qquad \text{(F11)}
$$

Gaussian curvature

Three-dimensional surfaces are classified into three groups according to their Gaussian curvature: elliptical surfaces (e.g. domes) have a positive Gaussian curvature (synclastic curvature) as the two principal curvatures are on the same side of the surface. On the other hand, hyperbolic surfaces (saddle surfaces) have a

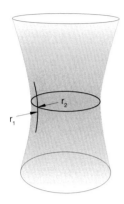

8

Gaussian curvature K

$-3 \cdot 10^{-4}$ $-20 \cdot 10^{-4}$

9

negative Gaussian curvature (anticlastic curvature); the two principal curvatures are on different sides of the surface. The third group form parabolic surfaces (e.g. cylinder and conical shells), which have a curvature $K = 0$ as these surfaces are curved in only one direction. These singly curved surfaces are always developable, unlike doubly curved surfaces.

The surface of the one-sheeted hyperboloid is anticlastic, having principal curvatures of opposite sign at a given point. The Gaussian curvature K is therefore always negative over the whole of the surface. It is defined as the product of the two principal curvatures k_1 and k_2:

$$K = k_1 \, k_2 = \frac{1}{r_1} \, \frac{1}{r_2} \qquad\qquad \text{(F 12)}$$

The Gaussian curvature [6] at a specific height z can be calculated as:

$$K(x, y, z) = - \frac{c^6}{(c^4 + a^2 z^2 + c^2 z^2)^2} \qquad\qquad \text{(F 13)}$$

This gives the maximum Gaussian curvature at the waist as $K = -c^2$. As can be seen from Fig. 9, as the necking becomes more severe the maximum Gaussian curvature increases but becomes more concentrated at the waist and decreases again towards the edges.

Geometry of hyperbolic lattice structures
The geometry and shape of hyperbolic lattice structures can be determined from the general mathematical relationships discussed in the previous section.

Fundamental relationships
The starting point for this determination is two parallel circles (or ellipses) arranged one above the other at a distance H apart. Each of these circles is divided by n pairs of vertical members into n equal sections. The sections of arc subtend the central angle ψ at the centre of the circle (Fig. 10).

$$\psi = 360°/n \qquad\qquad \text{(F 14)}$$

A family of straight vertical members is now arranged between the two circles (or ellipses) in such a way that the start and end points of the members in plan are each displaced by a rotation angle φ. A second family of straight lines rotated by the angle φ in the opposite direction is added. If the rotation angle is φ a multiple of the central angle ψ then the divisions of the two circles are directly above one another in plan.

$$\varphi = k\psi/2 \qquad\qquad \text{(F 15)}$$
$$K_\varphi = \varphi/\psi \qquad\qquad \text{(F 16)}$$
$$n_{SP} = 2K_\varphi - 1 \qquad\qquad \text{(F 17)}$$

The larger the rotation angle, the further the start and end points of a line are turned with respect to one another, and the more pronounced is the characteristic necking of the hyperboloid when viewed in elevation (Fig. 11, p. 30). The maximum necking is achieved at rotation angle of 180°: All the straight members now intersect one another at one point in the middle – the hyperboloid has become a double cone.

As the rotational displacement increases, the necking becomes more pronounced and the number of intersection points of the straight members n_{SP} increases, dividing the hyperboloid into n_{SP}-1 sections.

The basic geometry of the hyperbolic lattice structure can be determined from five independent shape parameters:

• Bottom radius R_U
• Top radius R_O
• Height H
• Number of members n
• Rotation angle φ

In the following, the ratio of the top to the bottom radius is defined as the shape parameter K_F.

Position of coordinates

The coordinates of the lattice structure can be deduced from the linear equations (Fig. 12, p. 31). The start and end points of a generatrix take the following form in vector notation:

$$A = \begin{bmatrix} R_u \cos \psi_A \\ R_u \sin \psi_A \\ 0 \end{bmatrix}$$

(F 18)

$$B = \begin{bmatrix} R_o \cos (\psi_A + \varphi) \\ R_o \sin (\psi_A + \varphi) \\ H \end{bmatrix}$$

The generatrix E of the hyperboloid surface has therefore the form:

$$E_{Hyp} = \begin{bmatrix} R_u \cos \psi_A \\ R_u \sin \psi_A \\ 0 \end{bmatrix} + \lambda \begin{bmatrix} R_o \cos (\psi_A + \varphi) - R_u \cos \psi_A \\ R_o \sin (\psi_A + \varphi) - R_u \sin \psi_A \\ H \end{bmatrix}$$

(F 19)

where $\lambda \in \{0,1\}$ and $\psi_A \in \{0°, 360°\}$

To obtain the equations of the two counter-running member families, the end point is rotated by $+\varphi$ or $-\varphi$ with respect to the start point. The individual members are defined by the multiple of the central angle ψ:

$$g_m = \begin{bmatrix} R_U \cos (m\psi) \\ R_U \sin (m\psi) \\ 0 \end{bmatrix} + \lambda \begin{bmatrix} R_O \cos (\varphi + m\psi) - R_U \cos (m\psi) \\ R_O \sin (\varphi + m\psi) - R_U \sin (m\psi) \\ H \end{bmatrix}$$

(F 20)

where $m \in N \cap \{0, 1, ..., n-1\}$

$$h_m = \begin{bmatrix} R_U \cos (m\psi) \\ R_U \sin (m\psi) \\ 0 \end{bmatrix} + \mu \begin{bmatrix} R_O \cos (-\varphi + m\psi) - R_U \cos (m\psi) \\ R_O \sin (-\varphi + m\psi) - R_U \sin (m\psi) \\ H \end{bmatrix}$$

(F 21)

where $m \in N \cap \{1, 2, ..., n\}$

8   Opposite principal curvatures of the one-sheeted hyperboloid
9   Increasing Gaussian curvature towards the waist of the one-sheeted hyperboloid
10 Central angle divisions for circular and elliptical plan forms

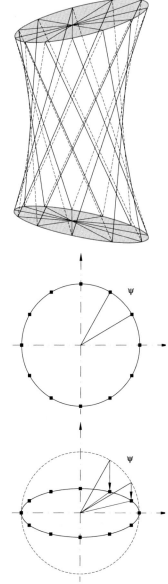

10

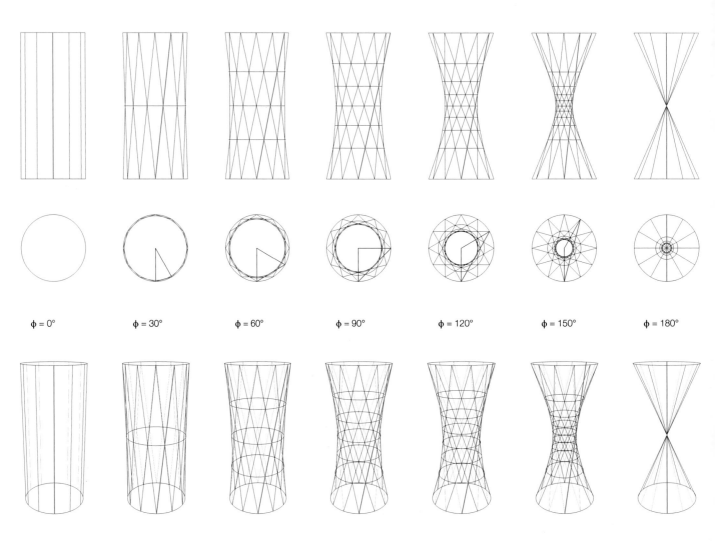

$\phi = 0°$ $\phi = 30°$ $\phi = 60°$ $\phi = 90°$ $\phi = 120°$ $\phi = 150°$ $\phi = 180°$

Geometry and form of hyperbolic lattice structures

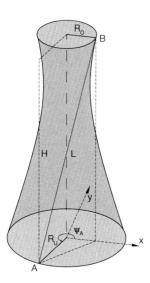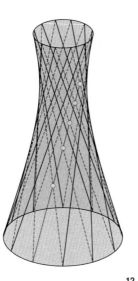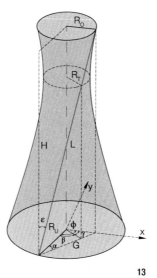

11 Influence of the rotation angle on the geometry in elevation, plan and axonometric projection
12 Position and number of intersection points
13 Position of the generatrix in space

12

13

In order to calculate the intersection points of the families of straight lines, the straight line in the first family for which m = 0 is determined. The intersection points shown in Fig. 13 are determined by equating F20 and F21. From the third line follows the equivalence of the linear parameters λ and μ. For the second line, the linear parameter μ is:

$$\mu_i = \frac{R_U \sin(i\psi)}{R_U \sin(i\psi) + R_O [\sin(\varphi) - \sin(i\psi - \varphi)]} \qquad \text{(F22)}$$

where $i \in N \cap \{1, 2, ..., n_{SP}\}$

As in the case of $i\psi = 180°$, the numerator as well as the denominator of the fraction become zero, the position of the intersection point can be determined using L'Hospital's rule [7]:

$$\mu_i = \frac{R_U \psi \cos(i\psi)}{-R_O \psi \cos(i\psi - \varphi) + R_U \psi \cos(i\psi)} \qquad \text{(F23)}$$

where $i \in N \cap \{1, 2, ..., n_{SP}\}$

Geometry of the frame mesh
All the geometric values and dimensions of the frame mesh can be deduced from the basic parameters listed above. Using simple relationships, it is possible to derive the length of the vertical member in the plan projection G, its true length L as well as the member inclination relative to the vertical axis ε (Fig. 13):

$$G = \sqrt{R_U^2 + R_O^2 - 2 R_U R_O \cos \varphi} \qquad \text{(F24)}$$

$$L = \sqrt{G^2 + H^2} \qquad \text{(F25)}$$

$$\varepsilon = \arctan(G/H) \qquad \text{(F26)}$$

In the plan triangle $R_O - G - R_U$, the angle α is determined from:

$$\alpha = \arcsin \cdot \left(\frac{R_O \sin \varphi}{G} \right) \qquad \text{(F27)}$$

And with this the waist radius R_T from:

$$R_T = R_U \sin \alpha \qquad \text{(F28)}$$

The height of the waste H_T comes from the member inclination as:

$$H_T = \frac{R_U \cos \alpha}{\tan \varepsilon} \qquad \text{(F29)}$$

The radius at any height h can be calculated from:

$$r(h) = \sqrt{R_U^2 + \left(\frac{h}{H} G\right)^2 - 2 R_U \left(\frac{h}{H} G\right) \cos \alpha} \qquad \text{(F30)}$$

Now the waist angle β introduced in plan. This angle expresses the aperture between the start point and the waist point of the generatrix:

$$\beta = \arccos \frac{R_T}{R_U} \text{ or } \beta = 90° - \alpha \qquad \text{(F31)}$$

If the top ring does not coincide with the waist, then R_O does not equal R_T, and the angular difference γ is calculated from the distance between the waist angle β and the rotation angle φ:

$$\varphi = \beta \pm \gamma \qquad \text{(F32)}$$

Structural analysis and calculation methods

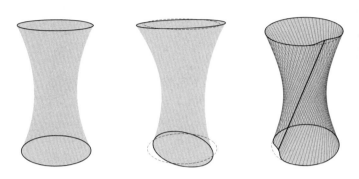

1

Hyperbolic lattice structures occupy a special position between reticulated shells and space frames. The type and size of mesh influence the structural behaviour and load transfer characteristics. The principal structural behaviour under vertical or horizontal loads can be easily derived from that of space frames. The basic methods of calculation used to determine the ultimate load are similar to the established methods used to analyse gridshell structures.

The problem of inextensional bending

All shell types with straight generatrices (ruled surfaces) can deform without strain. These include cylinders, cones, hyperboloids, conoids and hyperbolic paraboloids. The principles of this problem can be found mainly in older publications on shell structures such as Lajos Kollar's "Die dehnungslosen Formänderungen von Schalen", published in 1974 [1]. Inextensional bending means that the middle surface of a shell can deform without mobilising membrane stresses. In these circumstances, no strain takes place and shear deformation likewise remains at zero. Resistance to the deformation can only be provided by the secondary bending and torsional stiffnesses of the shell. Fig. 1 shows the two critical cases of inextensional bending of a one-sheeted hyperboloid: the ovalisation of the edges as well as the radial displacement of a straight line generatrix. With one-sheeted hyperboloids, the edges must be stiff enough in bending to prevent inextensional bending. [2]

Principal structural behaviour

The following paragraphs illustrate the load transfer behaviour of a single-storey hyperbolic lattice tower of a Shukhov-type design. For this example, horizontal and vertical point loads are applied to the top ring. The vertical member forces can be calculated quite simply from the geometry using the equations in the previous chapter.

Vertical load transfer

Vertical point loads applied to the top ends of the vertical members are transferred through the top ring into the vertical member force F_S (Fig. 2) and, because the vertical members are inclined, in their vertical plane, into the ring force N_O:

$$F_S = \frac{F}{\cos \varepsilon} \qquad \text{(F01)}$$

An important aspect influencing structural behaviour is the position of the waist or necking. Depending on the basic geometry, the hyperboloid has a physical waist or, in some cases, a notional waist located above the top ring (Fig. 3c). As the generatrices are tangential to the waist diameter in plan, the rotation angle φ can be divided into the angles β and γ, which determine the position of the necking in the horizontal projection:

$$\varphi = \beta \pm \gamma \qquad \text{(F02)}$$

The position of the waist has an important influence on the forces in the vertical members under vertical loading: if the tower has a necking, the straight members undercut the top ring in plan. This means a vertical load applied to the top member ends (e.g. the self-weight of a water tank) induces a horizontal tensile force in the top ring. The top ring acts as a tension ring and prevents the vertical members from falling outwards (Fig. 3a). However, if the waist lies above the top ring – i.e. the hyperboloid has no necking – then the vertical members lean on the top ring from outside: under the action of self-weight, the top ring is now subject to a compressive force (Fig. 3c). A third possibility is that γ equals zero and therefore the waist of the tower coincides with the top edge (Fig. 3b). In this limiting case, a pair-wise arrangement of vertical members results in no force in the top ring under self-weight. Two opposing vertical members form a straight line in plan – their horizontal forces cancel one another out at their upper ends. The ring forces can be calculated quickly from the geometric relationships. The horizontal component of the vertical members subjected to normal force F can be calculated from:

$$F_H = F_S \sin \varepsilon \qquad \text{(F03)}$$

Hyperbolic structures: Shukhov's lattice towers – forerunners of modern lightweight construction, First Edition. Matthias Beckh.
© 2015 John Wiley & Sons, Ltd. Published 2015 by John Wiley & Sons, Ltd.

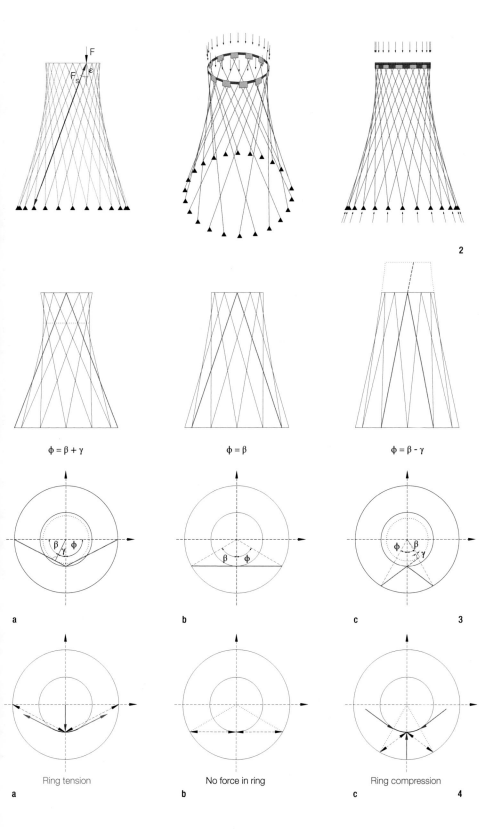

2

$\phi = \beta + \gamma$

$\phi = \beta$

$\phi = \beta - \gamma$

a

b

c

3

Ring tension

No force in ring

Ring compression

a

b

c

4

1 Characteristic inextensional bending deformation of the one-sheeted hyperboloid generated by a straight generatrix: ovalisation of the edges and parallel displacement of the straight members

2 Vertical load transfer

3 Relationship between the rotation angle and the position of the throat circle or "waist"
  a Waist below the top ring
  b Waist coincident with the top ring
  c Waist above the top ring

4 Normal forces on the top ring: ring tension (a), horizontal components of the vertical members cancel each other out (b), ring compression (c)

5 Resolution of the member force into normal and tangentially acting components on the top and bottom ring
6 Normal force in the top ring shown relative to the rotation angle φ using the example of the key geometric data of the water tower in Mykolaiv
7 Normal force distribution under a horizontal top load assuming the top edge is stiff in bending, compression and tension. Despite the presence of the intersection points, the normal forces in each vertical member are constant over the height of the tower. In this case, any intermediate rings are unloaded until the buckling loads of the vertical members are reached. (The intermediate rings are not shown in the drawing.)
8 Calculation of the vertical member forces under the action of a horizontal top load

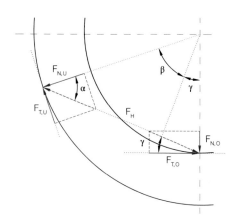

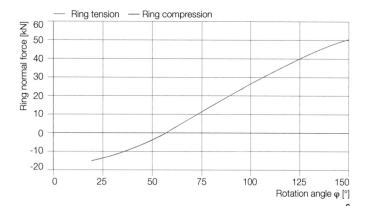

5

6

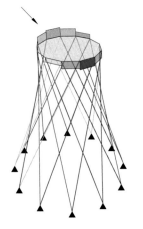
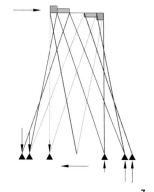

7

Using the angles α and γ, this horizontal component can be resolved at both ends of the vertical member into a component acting radially and a component acting tangentially on the ring (Fig. 5). The force in the ring can now be calculated primarily from the normally acting components. If these components acting perpendicularly to the ring are expressed as a combined uniformly distributed load, then the ring normal forces can be calculated simply using Barlow's formula. The tangential component adds to this force, but this component is secondary and alternates with the change in orientation of the incoming vertical members. The force components of the vertical members and the resulting normal forces in the top and bottom rings can be calculated from the following equations:

Top ring:

$$F_{N,O} = F_H \sin\gamma \qquad \text{(F 04)}$$
$$F_{T,O} = F_H \cos\gamma \qquad \text{(F 05)}$$
$$p_O = n\, F_H \sin\gamma / (\pi R_O) \qquad \text{(F 06)}$$
$$N_O = p_O R_O \pm 0.5\, F_H \cos\gamma \qquad \text{(F 07)}$$

Bottom ring:

$$F_{N,U} = F_H \cos\alpha \qquad \text{(F 08)}$$
$$F_{T,U} = F_H \sin\alpha \qquad \text{(F 09)}$$
$$p_U = n\, F_H \cos\alpha / (\pi R_U) \qquad \text{(F 10)}$$
$$N_U = p_U R_U \pm 0.5\, F_H \sin\alpha \qquad \text{(F 11)}$$

For example, for the geometry of the tower in Mykolaiv with a K_F-value of 1.83, a total height of 25.6 m and 48 vertical members, the normal forces in the top ring were calculated for various rotation angles. All the vertical members were loaded at the top end with a vertical force of 10 kN. Fig. 5 shows the flow of force in the top ring as the rotation angle is varied. The transition from compression to tension in this case takes place at $\varphi = 56.9°$; here the waist radius coincides with the radius of the top ring. The actual rotation angle φ of the tower is 82.5° and the top ring is therefore in tension under vertical loads.

Horizontal load transfer
Thanks to the regular construction of the lattice, it is also possible to calculate the forces in the vertical members under horizontal loads using geometric relationships. The load transfer of individual horizontal node forces can be derived from the general case of a horizontal top load.

Horizontal top load
If a horizontal load is applied to the top of the tower, the load is distributed by the top ring – in this case idealised as being stiff in bending, compression and tension – into the vertical members in accordance with their geometric stiffness. Each pair of vertical members meeting at the top edge forms a simple inclined bipod which is under uniform compression or tension (Fig. 7). The normal forces in each leg are unaffected by intersection points and remain constant over the height of the tower. Any intermediate rings (not shown in the drawing) carry no load in this case until the vertical members exceed their buckling limit.
Each bipod takes on a greater or lesser share of the external load, depending on the position of the line connecting the bearing points

Structural analysis and calculation methods

of its legs. The better this connecting line aligns with the effective line of action of the force, the greater are the forces in that pair of vertical members. The distribution of the forces can be determined from the geometric stiffnesses of the bipods. For this, the displacement method can be used. The stiffness with respect to displacement in the horizontal direction I_H is calculated from the sum of the member extensions and compressions in the direction of force. The latter, on the other hand, are calculated from the squares of the cosine components of the angle between the direction of the horizontal force H and the hypotenuse of the relevant triangle in plan. The angle δ between the members of the bipod and the line connecting the bearing points can be determined in the plan projection for a triangle beginning on the x-axis:

$$\delta = \frac{180° - 2 \cdot \varphi}{2} \qquad \text{(F 12)}$$

For the connecting lines of each pair of vertical members, the angle δ is calculated by adding or subtracting multiples of the central angle. The displacement stiffness of the flattened hyperbolic lattice structure is calculated as:

$$I_H = \sum_{K=0}^{n/2} \cos^2 (k \cdot \psi + \delta) = \cos^2 (k \cdot \psi - \delta) \qquad \text{(F 13)}$$

The proportion of the force H on each vertical member pair and the force component in the plane of the bipod $F_{S, tan}$ can be calculated from the ratio of the stiffness of a bipod to the displacement stiffness I_H:

$$F_{S, tan} = H \cdot \frac{\cos^2 (k \cdot \psi \pm \delta)}{I_H} \cdot \frac{1}{\cos (k \cdot \psi \pm \delta)} \qquad \text{(F 14)}$$

The force in the vertical member follows from the solid angles γ and ε (Fig. 8):

$$F_S = \frac{F_{S, tan}}{2} \cdot \frac{1}{\cos \gamma} \cdot \frac{1}{\sin \varepsilon} \qquad \text{(F 15)}$$

Therefore the position of the most highly loaded vertical member pair (bipod) depends on the rotation angle: the bipod that has the connecting line in plan with the smallest angle to the line of action of the force carries the greatest load.
In the case of a whole number of vertical member pairs and a rotation angle of 90°, the connecting line coincides with the line of action. This vertical member pair, whose bearing points lie on the x-axis carries the greatest proportion of the applied load and the greatest support reaction forces. (Fig. 9, p. 36). With smaller or larger rotation angles, the connecting line deviates further and further from the line of action of the applied external force. In this case, the most heavily loaded vertical member pair – and conse- quently the one with the maximum support reaction force – is the pair whose connecting line makes the smallest angle to the line of action of the external force, or the one that comes the closest to it. This is made clear in Fig. 9, which shows the most heavily loaded

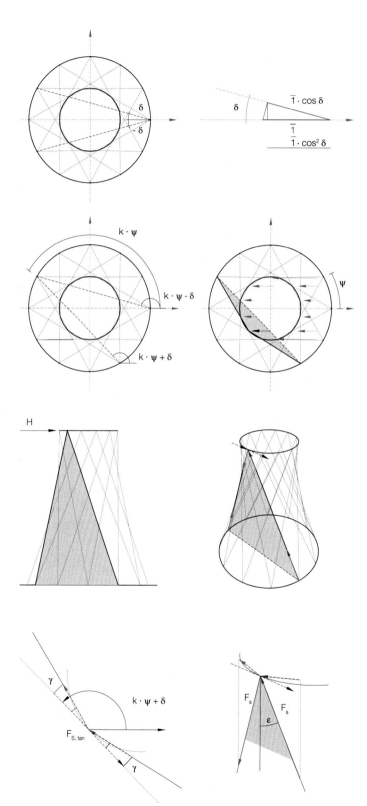

8

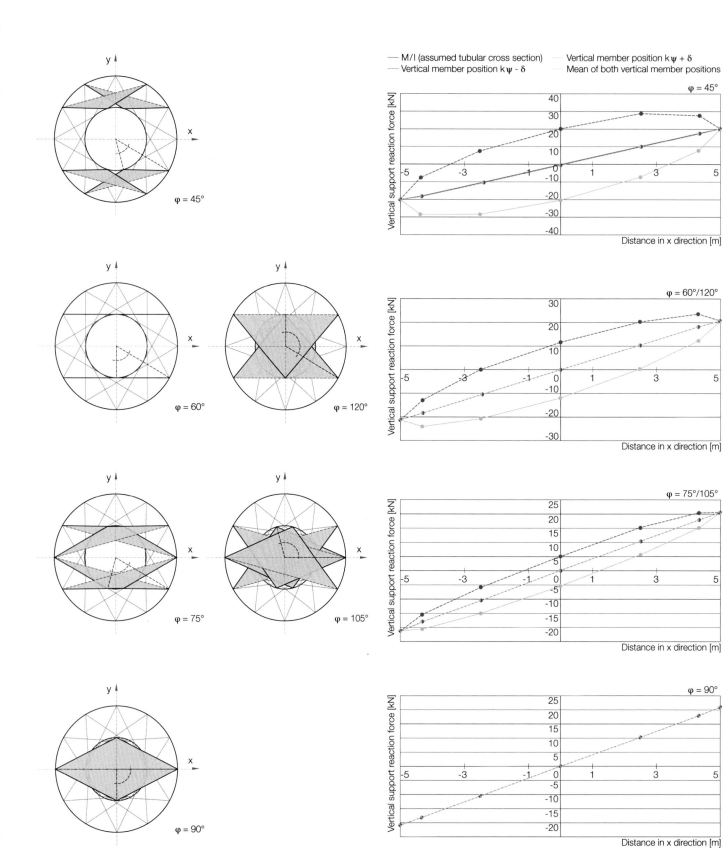

vertical member pair hatched in grey. In Fig. 10, the vertical support reactions for a horizontal top load of 100 kN are shown for various rotation angles ($n = 12$; $K_F = 2.0$; $H = 12.5$ m; $R_U = 5.0$ m). The mean values of the vertical reaction forces of the vertical members coming together at the base are linear over the cross section. They correspond with the results obtained assuming a shear-stiff tubular cross section (see "Calculations for Vladimir G. Shukhov's lattice towers", p. 70ff.) – the graphs are coincident. However, the forces in the pairs of vertical members meeting at the support points are quite different. While in one of the vertical member positions ($k \psi - \delta$), the member and support reaction forces lie above the linearly calculated line, the forces in the other vertical member position ($k \psi + \delta$) are below it by the same amount. Although the distance of the vertical members from the zero line is obviously the same at the support, the member and support reaction forces are different, depending on the orientation of the vertical members in three-dimensional space. A calculation of the vertical member forces assuming a shear-stiff tubular cross section therefore delivers incorrect member forces, as can be seen from the results in Fig. 10. For example, the vertical support reactions for a rotation angle of 45° are 33 % above the value calculated from the tubular cross section hypothesis, while for a rotation angle of 60° it is still 13 % above. Unlike the structural behaviour of a shear-stiff lattice tube, in the case presented here the vertical members at the extreme edge of the whole structure are not always the most heavily loaded.

Horizontal node loads

If horizontal point loads are applied at every node, the intermediate rings distribute the loads between the vertical members. The rings on the windward side are loaded in compression, while those on the leeward side are in tension (Fig. 11 a). The behaviour is similar to the case of the horizontal top load, but with vertically spaced layers. The vertical members adjacent to the nodes attract proportionately more or less load depending on their geometric stiffness, the normal force increases incrementally at every intersection (Fig. 11 b and c).

If the intermediate rings are not arranged at the nodes as shown in Fig. 11 but are positioned between nodes (the Shukhov form of construction), then the vertical members are also loaded in bending (this is given further consideration in the chapters "Relationships between form and structural behaviour", p. 50ff. and "NiGRES tower on the Oka", p. 96ff.).

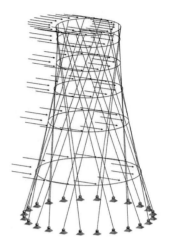

a

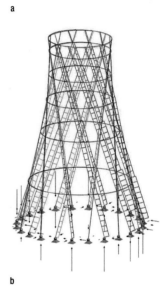

b

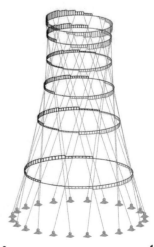

c

9 Rotation angle and the associated vertical member pair with the most heavily loaded bipods for a top load in the x-direction highlighted in light grey

10 Distribution of the vertical support reactions over the cross section for different rotation angles and a horizontal top load of 100 kN ($R_U = 5.0$ m; $K_F = 2.0$; $n = 12$; $H = 12.5$ m)

11 Effect of horizontal node loads: loading (a), normal force distribution and support reactions (b), normal force distribution in the intermediate rings, enlarged view (c)

11

Influence of torsion

If the straight line generatrices of the hyperboloid are manufactured out of steel angle sections, as is the case with many of Shukhov's towers, then the sections are twisted over their complete length (Fig. 12). As the sections are typically tangential to the bottom and top rings, the angle the vertical members are turned through is identical to the rotation angle φ of the generatrices of the hyperboloid. For warp-free open angles or U-sections, the torsional moment resulting from the twisting of the vertical members and the associated torsional stresses can be calculated in accordance with Saint Venant's principle from:

$$M_T = \varphi \; \frac{\pi}{180} \; \frac{1}{L} \; G \, I_T \qquad\qquad \text{(F 16)}$$

$$\tau = \frac{M_T \, t}{I_T} = \varphi \; \frac{\pi}{180} \; \frac{1}{L} \; G \, t \qquad\qquad \text{(F 17)}$$

Shukhov's first tower, which was built for the All-Russia Exhibition in Nizhny Novgorod, has the following values: unrestrained member length in space: $L = 2628$ cm, $\varphi = 90°$, member cross section $\sum 7.5/7.5/1.0$ cm with $I_T = 5$ cm^3. Applying the above equations gives a constant member torsional moment M_T of 0.24 kNm and a resulting shear stress of 4.84 kN/cm^2. The stresses arising from torsion were omitted from the calculations as their influence is relatively small. It is currently not possible to say whether the vertical members were fixed into their twisted position or were deformed plastically at the fabricator's factory.

Theoretical principles for determining ultimate load capacity

In addition to their basic structural behaviour, lattices and shell structures under compressive load must also be checked for their buckling stability. Due to most of the members having very slender dimensions, the ultimate load capacity of lattice structures is more often determined by stability than by stress failures. Since the start of the 1990s, a number of publications have appeared which deal with the stability and the ultimate load

capacity of lattice- and gridshells. In earlier works, the common approach to problems of stability is from the point of view of a continuum shell. [3] This has in theory the advantage that it can draw on extensive literature about shell analysis. However, the methods used by the various authors to flatten the lattice structure diverge so greatly, as do the results achieved, that this methodology has never really caught on. The next possibility is to model the lattice shells as complex frames. This approach has become the default, thanks to the increase in available computing capacity.

Pioneering publications have appeared from the University of Stuttgart in recent years. For example, the publication *"Untersuchungen zum Tragverhalten von Netzkuppeln"* describes the basic problem and the main method used by consulting engineers schlaich bergermann und partner for the design of gridshells. [4] Several publications on the same theme build upon this to refine and extend the method. [5] Also worthy of particular mention is the dissertation by Jürgen Graf on the design of translation gridshells [6] as well as works which transfer the developed method of design of gridshells to free-form surfaces such as those of the new trade fair building Milan [7].

The main method of determining the ultimate load capacity in the calculations and parametric studies discussed here focuses on these works and transfers the established methods to hyperbolic lattice structures with their comparatively high overall stiffness. The following sections set out the theoretical principles for the structural analyses.

Stress and stability failures

The assessment of the structural behaviour of gridshells or spatially complex frames is usually performed using load-displacement diagrams (L–D diagrams) in which the load F is increased until failure occurs when the ultimate load F_{crit} is reached. The load-displacement diagram may relate to stress or stability failure, depending on its form and extent. In addition, local stability failures may be caused by buckling of individual members, which may define the achievable ultimate load capacity (Fig. 13). [8]

12   Torsion of the angle profiles in the water tower for the All-Russia Exhibition in Nizhny Novgorod (RUS) 1896, which stands today in Polibino

Structural analysis and calculation methods

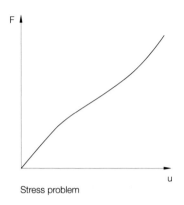

Stress problem

a

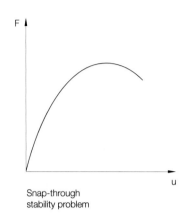

Snap-through
stability problem

b

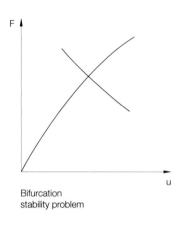

Bifurcation
stability problem

c

13

Stress-controlled failure
If a displacement state can be assigned to every load level, then
the problem is one relating to stress. According to Ekkehard
Ramm [9], three characteristic cases can be described by referring
to stress-strain curves (Fig. 14):
a System with increasing stiffness, e.g. a membrane-supported plate
   under transverse load
b System with decreasing stiffness, e.g. a stretched rubber band
c Combination of a and b, e.g. a very flat, not-too-thin spherical
   shell

In the case of stress problems, reaching the failure stress of the
material determines the ultimate load of the structure.

Stability-controlled failure
In contrast to the stress problem, with the stability problem a load
level can have more than one equilibrium state. For reticulated shells,
there are three different types of stability failures:

Buckling of a member
In this case, the lattice- or gridshell fails locally by the buckling of a single, heavily loaded vertical member, e.g. near a support or one with a
concentrated load.

Node snap-through (local stability)
Node snap-through describes the failure of parts of the structure
due to individual nodes snapping through out of the centroidal
plane of the shell and causing at least some parts of the system
to become unstable. The structure loses stiffness in these areas
with the result that its structural behaviour alters and a redistribution of the load flow occurs, which may allow further increases
in load.
Alternative models for investigating local instability can be found in
some publications. [10] They involve removing areas of the lattice
shell and investigating them separately. Difficulties arise, however,
when trying to model the support conditions. The stiffness of the
rest of the structure is either ignored, which means the model is

less representative of reality, or the effort involved in accurately
modelling the edge stiffnesses is so immense that the advantage
of analysing these partial systems is much reduced.

Global buckling (global stability)
Global buckling means the structure suffers a large-scale stability
failure.

Linear structural behaviour and stability analysis
The equilibrium of a discretised system can be described
generally using:

$$\underline{\underline{K}}\,\underline{u} = \underline{F} \qquad\qquad (F\,18)$$

$\underline{\underline{K}}$ Stiffness matrix
$\underline{u}$ Vector of the node displacements
$\underline{F}$ Vector of the external loads

The equation system can be directly solved by transformation.
The node displacements can be calculated from:

$$\underline{u} = \underline{\underline{K}}^{-1}\,\underline{F} \qquad\qquad (F\,19)$$

The equilibrium relates to the undeformed system. As the system
stiffness depends on the state of deformation, the graph is linear
and the various load cases can be superimposed on one another.

Linear eigenvalue analysis
Stability failures of a structure occur when no unique solution can be
assigned to a specific load level on the load-displacement curve. This
happens when, in addition to the basic equilibrium state u_g, a further
neighbouring equilibrium state at the same load level and with the displacement u_n exists at the node:

$$\underline{\underline{K}} \cdot \underline{u}_g = \lambda \cdot F$$
$$\underline{\underline{K}} \cdot \underline{u}_n = \lambda \cdot F \qquad\qquad (F\,20)$$
$$\underline{\underline{K}} \cdot [\underline{u}_g - \underline{u}_n] = \underline{\underline{K}} \cdot \Delta\underline{u} = 0 \qquad\qquad (F\,21)$$

Structural analysis and calculation methods

Load-displacement characteristics

13 Characteristic load-displacement curves: stress problem (a), snap-through
   stability problem (b), bifurcation problem (c)
14 Load-displacement characteristics of systems with stress problems
   a System with increasing stiffness, e.g. a membrane-supported plate under
      transverse load
   b System with decreasing stiffness, e.g. a stretched rubber band
   c Combination of a and b, e.g. a very flat, not-too-thin shell, in the case of stress
      problems, reaching the material's ultimate strength determines the ultimate
      load of the structure.

14

The following applies for the non-trivial solution $\Delta u \ne 0$:

$$\det (\underline{K}) = 0 \qquad\qquad (F\,22)$$

For stability analyses, the stiffness matrix K must be extended by
components that reflect the basic deformation and the load state.
The geometric stiffness matrix is calculated for the equilibrium
conditions on the deflected system ¬ using the principle of virtual
displacements:

$$\underline{K} = \underline{K}_e + \underline{K}_u + \underline{K}_g \qquad\qquad (F\,23)$$

$\underline{K}_e$ Linear-elastic stiffness matrix
$\underline{K}_u$ Initial displacement matrix
$\underline{K}_g$ Geometric stiffness matrix

If the initial displacement state is ignored, the system stiffness
simplifies to:

$$\underline{K} = \underline{K}_e + \underline{K}_g \qquad\qquad (F\,24)$$

The eigenvalue problem is then expressed as:

$$[\underline{K}_e + \lambda \cdot \underline{K}_g] \cdot \underline{\Phi} = 0 \qquad\qquad (F\,25)$$

The solution of the equation supplies the critical load factor λ_k as the
lowest eigenvalue and therefore the critical load $F_{crit} = \lambda_k F$. The eigen-
vector Φ associated with λ_k determines the shape of the buckling
mode. This process is a classical linear stability analysis.

Non-linear structural behaviour and stability analysis
In the case of geometrically non-linear behaviour, the system stiffness
and the loads depend on the displacement state. Equation F 18 can
be extended to:

$$\underline{K}(\underline{u})\ \underline{u} = \underline{F}(\underline{u}) \qquad\qquad (F\,26)$$

As the two sides of the equation depend on the loading history,
the solution must be calculated in steps. Methods for solving
non-linear equation systems can be generally classified as
incremental, iterative and incremental-iterative. Incremental
methods increase the loads in stages and assume linear behaviour
for each load stage. Nevertheless, the total error arising from this
approximation builds up with the increase in the number of load
stages. Therefore a purely incremental method for solving
non-linear problems is not acceptable. In contrast to this, iterative
procedures apply the loading in one stage and determine the
displacement vector by iteration. However, this method often
requires many stages of iteration. Moreover it is impossible to
reproduce non-unique solutions, which depend on the loading
history (e.g. the snap-through problem with shell buckling). For
these reasons, incremental-iterative methods have become
established in practice. In these procedures, the loading is applied
in stages and after each load stage an iteration is performed using
a numerical method. [11]
In addition to the load-controlled methods, it may also be worthwhile
calculating the load-displacement curve as displacement-controlled or
using the arc-length method. Only by using this method can reliable
predictions be made about secondary buckling of shell structures. As
investigation of secondary buckling was not of interest to this book
and this method was not available in the software used, it will not be
discussed further here.

Newton-Raphson method
One of the most frequently used methods in finite element programs
for solving non-linear finite element equations is the load-controlled
Newton-Raphson method. [12] It is based on the array of the tangent
matrix K_T, which is determined from a Taylor series expansion. In contrast to other methods, it converges very quickly and is the basis of
many other equation solution methods. The equation system to be
solved is:

$$\underline{K}(\underline{u}) \cdot \underline{u} - \underline{F} = 0 \qquad\qquad (F\,27)$$

15 Newton-Raphson method: traditional (a) and
load-controlled (b)
16 Interaction relationship by Stanley Dunkerley
17 Effect of imperfection on the load capacity:
Euler column II (a), disk under transverse load (b),
circular cylindrical shell (c)

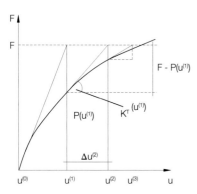

a

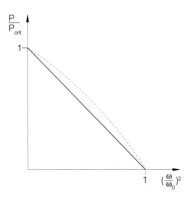

b

15

16

The linearisation of F27 is done using the expansion of a Taylor series in which higher order terms are ignored. For the kth iteration step:

$$(\underline{K}(\underline{u}) \cdot \underline{u})^{(k-1)} - \underline{F} + \left(\frac{d\,(\underline{K}(\underline{u}) \cdot \underline{u})}{d\,\underline{u}}\right)^{(k-1)} \cdot \Delta\underline{u}^k = \underline{0} \qquad \text{(F28)}$$

With the tangent stiffness

$$\underline{K}_T(\underline{u}^{(k-1)}) = \left(\frac{d\,(\underline{K}(\underline{u}) \cdot \underline{u})}{d\,\underline{u}}\right)^{(k-1)} \qquad \text{(F29)}$$

defined as the stiffness of the previous loading stage, one obtains:

$$\underline{K}_T(\underline{u}^{(k-1)}) \cdot \Delta\underline{u}^k = \underline{F} - \underline{P}(\underline{u}^{(k-1)}) \qquad \text{(F30)}$$

The right side of this equation gives the "out-of-balance forces", in other words the forces that take the system out of equilibrium in the $(k-1)^{th}$ iteration step. The vector of the internal forces in iteration step (k-1) is expressed by:

$$\underline{P}(\underline{u}^{(k-1)}) = (\underline{K}(\underline{u}) \cdot \underline{u})^{(k-1)} \qquad \text{(F31)}$$

The displacement increment can be calculated from F30 so that the new displacement state is given by:

$$\underline{u}^{(k)} = \underline{u}^{(k-1)} + \Delta\underline{u}^{(k)} \qquad \text{(F32)}$$

The iteration is repeated until the displacement increment Δu_k is less than a cut-off threshold. The traditional Newton-Raphson method is often used as an incremental-iterative method in practice and in the calculations that involve incremental increases in loading performed in the context of this book (Fig. 15b).

Non-linear stability analyses
In non-linear calculations, the snap-through point is reached when the tangent stiffness reaches the value zero, that is at the vertex of the L-D curve (Fig. 13b, p. 40). A further load increase cannot find equilibrium and the calculation cannot converge at this point. Therefore the load capacity is equal to the loading of the last converging increment step. The node displacements required to determine the buckling shape do not arise directly from the calculation, as is the case with linear buckling analyses. There is a direct relationship between the load and the eigenfrequency of a system – like the frequency of the sound from a string in a musical instrument depends of the tension in the string. In order to determine the shape of the buckling mode, the interaction relationship proposed by Stanley Dunkerley, which expresses the relationship of the eigenfrequency and the loading of systems whose vibrational and buckling modes are affine. Out of this comes the relationship between vibration mode and critical load (Fig. 16). The eigenfrequency under load from the last convergent iteration step is almost zero. The corresponding buckling mode is derived in this way from the modal analysis with the structure loaded to its load capacity.
Finding the branch points is more difficult. So as not to overlook them, an eigenvalue analysis must be done after each loading step

Structural analysis and calculation methods

using the current non-linear system stiffness. The intersection point of the L-D curve with the curve of the critical load from the eigenvalue analysis is the ultimate load in this case (Fig. 13c, p. 40).

Imperfections

In practice, structures can never be built to achieve their idealised design geometry. Deviations between the completed structure and its specified dimensions and conditions are described as imperfections. These deviations can be classified as:
• Geometric imperfections
• Structural imperfections

Geometric imperfections include deviations from the specified geometry and unplanned eccentricities of the point of application of the load. Structural imperfections include internal stresses caused by rolling, welding or material inhomogeneities. According to DIN 18800, equivalent geometric imperfections that simulate the geometric and structural imperfections are to be used in the design of a structure. Equivalent geometric imperfections for bars, frames, skeletal and arched structures can be found in DIN 18800 Steel structures – Part 2: Stability – Safety against buckling of linear members and frames. In accordance with DIN 18800-2 [13], the corresponding imperfections are applied in such a way that they are the best possible match for the deformed shape associated with the lowest buckling eigenvalue. They are also to be applied in the most unfavourable direction. For theoretically straight bars and frames, DIN 18800-2 makes a distinction between initial curvatures and initial twists. However, this distinction cannot be transferred to the consideration of the hyperbolic lattice structures examined in this book, because of their spatial complexity. The fourth part of the standard, "Stability – Analysis of safety against buckling of shells", cannot be transferred to lattice shells because it does not work with equivalent imperfections, but with reduced buckling stresses for the continuum shells. For hyperbolic lattice structures, which are in the boundary zone between shells and frames, it is therefore not a trivial task to find suitable equivalent imperfections.

Choice and size of imperfection shape

How much the load capacity is reduced by the presence of imperfections generally cannot be accurately predicted as this depends on the original geometry of the lattice structure or shell and the type and size of the imperfection shape. Therefore it is necessary to find the imperfection that leads to the lowest load capacity of the system (Fig. 17). In contrast to the case of columns, with shell structures the determinant imperfection shape is not necessarily aff-ine with respect to the first buckling eigenmode. The search for and selection of the most unfavourable imperfection shape therefore becomes particularly important.

The following method has become established for the analysis of lattice shells: normally the buckling eigenmodes associated with the lowest eigenvalue, or a combination of this and higher eigenval-ues is used as the determinant imperfection shape. If this displace-ment is now imposed at the appropriate scale, the system is forced to deviate into its weakest position. Less frequently, in instances where the first eigenvalues are close to one another, it can also be

— Ideal structure   ---- Imperfect structure

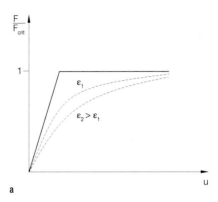

a

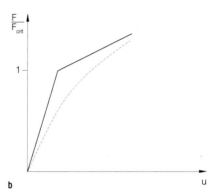

b

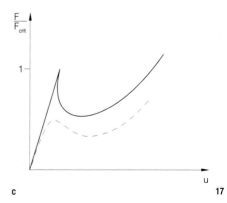

c   17

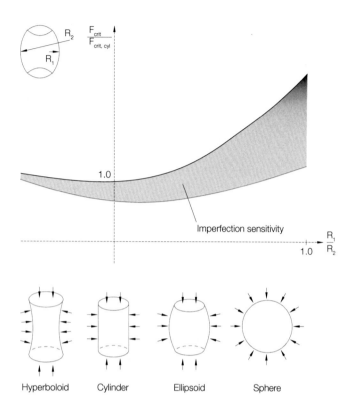

Hyperboloid   Cylinder   Ellipsoid   Sphere

18

18 Imperfection sensitivity of different shell structures under uniform external
   pressure
19 Three variants of one-sheeted hyperboloids with different meshes (a–c)
20 Schematic representation of the load transfer of horizontal node forces acting
   on the lattice
   a Transfer of load primarily into the rings and distribution into the verticals in
     accordance with geometric stiffness
   b Loading of the verticals from bending moments
   c Transfer of the load by normal forces into the rings and verticals

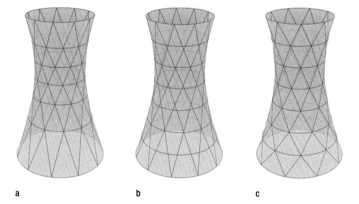

a   b   c

19

the case that higher buckling eigenmodes or combinations of
these lead to lower load capacities. [14] On the other hand, earlier
investigations of the stability behaviour of lattice shells that used
imperfection shapes based on deformation shapes significantly
overestimate the load capacity.
It would seem the best thing to do is to base the scaling on the
recommended values in DIN 18800-2. If the equivalent geometric
imperfections for fixed-ended arches are taken as a guide, then
this results in values of L/400 in the plane of the arch and L/200 at
right angles to the plane of the arch for cross sections of buckling
curve c (angle profiles, solid sections). In the case of loads in three
dimensions, the imperfections are applied in only one direction,
namely the most unfavourable. In some literature, the maximum
displacement is specified as 50 mm. It can be assumed from this
that an imperfection of this size can be seen with the naked eye and
therefore would be avoidable. [15] In the case of lattice structures
with straight generatrices, such as the hyperbolic structures in this
book, this hypothesis appears reasonable. The buckling eigenmode
is normally scaled with the help of the maximum norm and applied to
the system without imposing stresses.

The right choice of imperfection size is mainly determined by the
failure mode. If global stability failure takes place with large-scale
wave deformation patterns, then it is sensible to define the maximum value as a fraction of the span. For snap-through problems,
on the other hand, in which the deformations are very large but
locally limited in area, span ratios would lead to unrealistically large
deviations in position of individual bar elements. In this case no
scaling is made. For safety reasons, the starting point should
always be global stability failures. If local stability failures emerge
from the analysis, it can be repeated in a second step with reduced
imperfections. [16] According to DIN 18800-2, the stiffness of
systems at a higher risk of stability failure must be reduced by
dividing Young's modulus by $\gamma_M = 1.1$.

Relationship between shape and imperfection sensitivity
Imperfection sensitivity, the effect of deviations in shape on
the load capacity of the shell, depends basically on the type of
Gaussian curvature K. The report "Instability Behaviour of Single
Layer Reticulated Shells" describes imperfections sensitivity of four
different shell types related to the ratio of their principal curvature
radii (R_1/R_2). [17]
The report shows the ultimate loads F_{crit} of the shell under uniform
external pressure with respect to the cylinder shell $F_{crit, cyl}$. It shows
that, for shells with a positive Gaussian curvature, the ultimate
load F_{crit} for the perfect geometry certainly increases steeply, but
the imperfection sensitivity also increases overproportionately
(Fig. 18). On the other hand, for shells with negative Gaussian
curvature, such as the one-sheeted hyperboloid, although the
increase of the load capacity for the perfect geometry is
comparatively moderate, the reduction in load capacity caused
by imperfections is very small with increasing negative curvature,
with the result that the load capacities of the imperfect geometries
of the same proportions are once again similar. The secondary
curvature in the case of anticlastic surfaces produces less
pronounced buckling behaviour.

Parametric studies on differently meshed hyperboloids

The following three mesh variants were used to investigate structural behaviour: intermediate rings at the intersection points, the construction used by Vladimir Shukhov, and the reticulated shell (Fig. 19).

Intermediate rings at the intersection points (variant 1)

In this first variant, the intermediate rings are placed at the intersection points of the straight members. In addition to the top and bottom rings, the intermediate rings on this variant must be flexurally stiff in order to prevent kinematic movement of the structure (Fig. 20a). Horizontal point loads applied to the node points are distributed by the intermediate rings according to their stiffnesses to the triangular meshes below.

Construction used by Vladimir G. Shukhov (variant 2)

In the construction used by Shukhov, the intermediate rings are positioned at equal centres vertically up the height, independent of the intersection point positions (Fig. 20b). In addition to the top, bottom and intermediate rings being flexurally stiff, here the vertical members must also be flexurally stiff in order to prevent kinematic movement of the structure.

Reticulated shell (variant 3)

In the third variant the vertical members are not arranged along the generatrices of the hyperboloid. The intermediate rings are evenly distributed over the height and the intersection points are likewise evenly spaced from the bottom. As a result, there is a change in inclination at each node between the adjoining members (Fig. 20c). The double curvature of the surface of the one-sheeted hyperboloid is also present in the resolved mesh structure. The variant can also be made with vertical members that are pin-jointed together by connections at their ends that allow this change in inclination. In this case, only the rings at the top and bottom of the structure need to be flexurally stiff in order to prevent it from undergoing unwanted inextensional bending.

Principles of the parametric studies

Comprehensive parametric studies were carried out to gain a deeper knowledge of the influence of different design parameters. The results are discussed in the chapter "Relationships between form and structural behaviour" (p. 50ff.). The next section discusses the modelling of the three variants.

Finite Element Analysis

Ansys, the software used in batch mode for the study, allows the calculations for parameter studies to be controlled with text files written in the program's own object programming language Ansys Parametric Design Language (APDL). The variation of design parameters can be controlled through this combination of parametrised input and loop operations.

Beam type 188 linear elements were used to model the lattice structure. These elements are based on the Timoshenko beam theory and therefore take into account the influence of shear deformations. The elements are defined in three dimensional space by their start and end points i and j, which have three degrees of

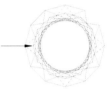

a

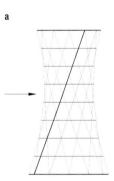
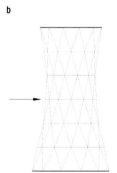

b

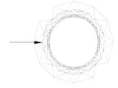
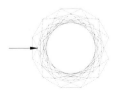

c

20

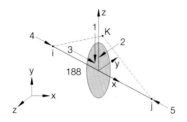

21

21 Beam element type 188 and cross section orientation
22 Vertical member cross sections and arrangement of the three investigated variants
23 Multilayer construction, arrangement of the joints at the inner edge of the intermediate ring
24 Lattice structure model in Ansys, arrangement of the joints at the inner edge of the intermediate ring
25 Comparison of the linear and non-linear ultimate load investigations based on the example of variant 2 ($K_F = 1.0$; IR = 10; n = 24; $\varphi = 90°$)

	Variant 1	Variant 2	Variant 3
Cross section of the vertical members [mm]	L120/120/12	L120/120/12	Solid profile Ø 60
Cross section of the intermediate ring [mm]	L 80/80/10	L 80/80/10	Solid profile Ø 60
Cross section of the top ring [mm]	2≈ L120/120/12	2≈ L120/120/12	Solid profile Ø 60
Arrangement of the vertical members	Multilayer	Multilayer	In one plane
Support at base	Fixed	Fixed	Fixed

22

translational and rotational freedom. An additional third node k defines the direction of the z-axis and can, for example, model any twisting of the cross section over the length of the vertical member (Fig. 21). The geometry and position of the individual node points were calculated using linear equations from "Geometry of the lattice mesh" (p. 31). The numerical calculations for the physical and geometric behaviour, both linear and non-linear, were carried out using Ansys. The non-linear calculations were load-controlled; the loading being applied in steps of 1/1000 of the linearly calculated load capacity. The design of the vertical members is based on the elastic-elastic method in accordance with DIN 18800-1.

Modelling and member arrangement
The finite element models for the three different variants are programmed to allow them to be easily modified for the design parameters as required by the geometry.

Dimensions
The height was set at a constant 25 m for all models studied. As a rule, the bottom ring radius is 5 m, the top ring radius varies. The ratio of the bottom and top radii was defined as the proportionality factor K_F. Likewise the number of vertical members n, rotation angle φ and the number of intermediate rings were varied.

Material
Structural steel grade S 235 was used for modelling the cross sections. At least the earlier towers by Shukhov used steel types with proven properties equivalent to those of S 235. A linear-elastic material was used for the results listed here:

$E = 210,000$ N/mm²
$\nu = 0.3$
$\rho = 7.85 \approx 10^{-6}$ kg/mm³

Vertical member cross sections
Sections similar to those used by Shukhov on comparable water towers were used to model the first and second mesh types. Because of the continuity of the vertical members, the multilayered aspect of the construction was taken into account with these types. For the modelling of the third mesh type on the other hand, solid circular profiles with a diameter of 60 mm were used in a single layer and pin-jointed at the node points.

Degrees of freedom
In all three variants, the vertical members were built-in at the base and connected with a flexurally stiff connection to the top ring (Fig. 25).

Variants 1 and 2
Like the design of the node points adopted by Shukhov, the verticals at the intersection points are connected by a flexurally stiff connection. At the intersection points, the connection of the eccentrically positioned vertical members is by fixed-ended vertical members made from a solid circular section (diameter 60 mm). These also form the transition between the vertical members to the intermediate rings on the inside face of the structure. The connection of the elements projecting from the vertical member is pin-jointed at the

intermediate ring – with the Shukhov towers, the intermediate ring is usually fastened to the projecting bracket with a bolt (Fig. 23).

Variant 3
The vertical members and node points are connected by a flexurally stiff connection. An investigation of the fixed-ended effect at the member ends is discussed in Fig. 23 (p. 59).

Load assumptions
The models were investigated according to their stability behaviour separately under vertical load and a combination of vertical and horizontal loads.
For the investigation of purely vertical loads, a top load of 100 kN was applied to the system and evenly distributed on the top vertical member ends. The calculations took no account of the self-weight of the structure because comparative calculations showed that the error resulting from this depended on the geometry and was 1.3 % in the extreme case, but on average was less than 0.5 % and therefore could be ignored. [18]
As with the linear and non-linear stability analyses, all the loads were scaled with the eigenvalue, while the horizontal influences were in proportion to the vertical loads. For the investigation of the combined vertical-horizontal load case combinations, the horizontal load was therefore defined as 5 % of the vertical load and likewise distributed on the member ends at the top ring. This corresponds with a tower used as an example for the study of the effects of wind force on the water tank. Stability failures were exclusively determinant for the vertical load capacity; stress problems on the other hand occur with slender towers under a combined vertical and horizontal loads and in these circumstances were determinant for the load capacity.

Choice of imperfections in determining the load capacity
The analyses used linear and non-linear load capacity calcul-ation methods, each done in two stages on perfect and imperfect geometries.
In choosing the imperfection shape, the effects of different buckling eigenmodes were investigated. It showed that the first eigenform always leads to the lowest load capacity. The second eigenform is usually the first eigenform turned through 90° and therefore leads to the same load capacity. Imperfection shapes that relate to the higher buckling eigenmodes lead without exception to larger load capacities.
In the linear buckling analysis, the load capacity was reduced by the application of imperfections of 5 to 10 % compared with the perfect geometry. The linear analyses were used to estimate quickly the effects on the load capacity in cases with a large number of different shape parameters. In addition, non-linear, load-controlled load capacity analyses were used for selected areas and for the water towers built by Shukhov. After the calculation of the load capacity of the perfect geometry, the determination of the corresponding buckling eigenmode was done using a modal analysis of the structure loaded to capacity; this corresponds to the interaction relationship by Stanley Dunkerley of the first buckling eigenmode. The scaled deformation was imposed in the linear case on the perfect geometry. The actual load capacity of the imperfect structure was determined in a second step.

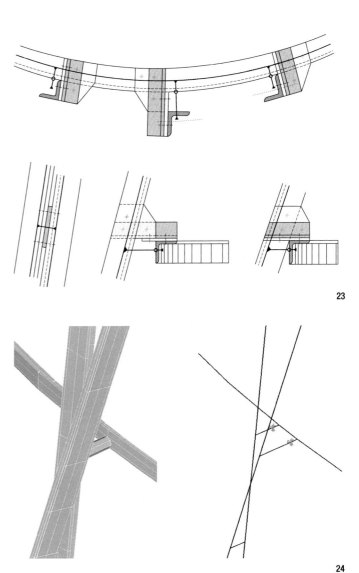

23

24

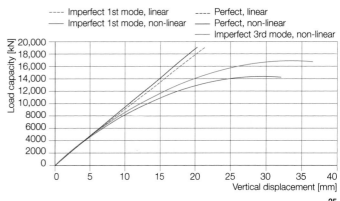

25

Structural analysis and calculation methods

Comparing the linear and non-linear analyses shows that the load capacity of the perfect geometry is the same in both cases (Fig. 25, p. 47); the curves overlay upon one another exactly. However, the non-linear load capacity is much smaller with imperfect geometry. While the reduction using the linear analysis is a maximum of 10 %, with most lying well below this value, the reduction using non-linear analysis is generally around 25 %.

Calculation of load capacity

The calculations are based on an elastic material model. The term "load capacity" in this book always refers to the elastic ultimate load without taking any increase in load due to load redistribution in the plastic zone into account. The calculation of the determinant elastic load capacity for the stability case is performed in the following steps for both the linear and non-linear analyses:

- Modelling the perfect structure with the finite element program Ansys
- Calculation of the buckling load P_{k1} assuming linear behaviour or using the non-linear load-displacement curve up to the ultimate load P_{k1}
- Calculation of the eigenvalues and the eigenmodes of the loaded structure
- Standardising and scaling of the buckling eigenmodes; normally used in the first mode and scaled to 50 mm
- Creation of the imperfect structure
- Calculation of the buckling load P_{k2} of the imperfect structure assuming linear behaviour or using the non-linear load-displacement curve up to the load capacity P_{k2}
- Calculation of the eigenform directly assuming linear behaviour using linear analysis or with the help of modal analysis of the last converging load step

In order to minimise the time required to process the large number of calculations for this, a linear load capacity analysis was used in most cases. Non-linear analyses were used for selected areas of the parametric study and for the load capacity calculations for the built Shukhov towers.

26 NiGRES tower on the Oka, transition from the third to the fourth segment, Dzerzhinsk (RUS) 1929

Relationships between form and structural behaviour

This chapter presents the results of the parametric studies carried out to examine the influences of each shape parameter on structural behaviour. The theoretical and computational principles were described in the section "Principles of the parametric studies" (p. 45ff.). Some of the results are based on the master's thesis on the influence of shape parameters on the structural behaviour of hyperbolic structures [1], which was published earlier in an article on the shape and structural behaviour of hyperbolic lattice towers [2].

Comparison of circular cylindrical shells and hyperboloids of rotation

In order to be able to evaluate the stability of hyperbolic lattice structures, simple continuum models of cylindrical shells are compared with three different hyperboloids of rotation. The qualitative investigation is intended to allow conclusions to be drawn about the suitability of one-sheeted hyperboloids for transferring vertical loads. The four different shells have a wall thickness of 50 mm and were modelled with no special stiffening of the top edge. The models are 25 m high, the top and bottom radii are both 5 m; the supports at the base are pinned.

The critical load factors were first determined linearly for a vertical top load of 100 kN. The results in Fig. 1 show that the intact cylindrical shell has the highest load capacities. If the models suffer from imperfections, the anticlastic curvature of the hyperboloids makes a noticeable positive contribution. The reduction of the load capacity becomes smaller with increasing negative Gaussian curvature. With moderate necking, the load capacities of the imperfect hyperboloids are greater than those of the imperfect cylinder shells, as the example hyperboloid with $\varphi = 60°$ shows. Because of the more favourable stability of anticlastic shell structures – investigated and demonstrated, for example, by Victor Gioncu and Nicolae Balut (Fig. 18, p. 44) [3] – natural-draught cooling towers are designed as one-sheeted hyperboloids. The reduced risk of buckling means that thinner wall thicknesses can be used than with cylindrical or truncated cone shells.

Mesh variant 1: Intermediate rings at intersection points

With this mesh variant, the intermediate rings (IR) are always at the level of the intersection points of the generatrices. In contrast to the Shukhov-built variant 2, the dependencies are clear to see and not overlain by various other effects.

Influence of the rotation angle φ

For the rotation angle φ of the first mesh type, multiples of the central angle ψ were used so that each vertical member end always met another at the top edge. Fig. 5 shows an overview of the buckling modes of the first mesh type, in which the intermediate rings are positioned at the heights of the intersection points of the generatrices. With a small rotation angle ($\varphi = 30°$), the unrestrained vertical member lengths are very large because of the low number of intersection points – and therefore the low number of horizontal rings. Buckling of individual vertical members about their weak principal axis determines the load capacity here. With larger rotation angles, the number of intersection points increases and the unrestrained span lengths decrease: the structure becomes stiffer. Failure is now much less local and takes place over a wider area. The optimum range for K_F-values lies between 1.0 and 2.0 for rotation angles between 90 and 120°; here the load capacities are approximately four times as high as with very small or very large rotation angles. In the case of still larger rotation angles, the number of intersection points continues to increase, but because these are concentrated towards the middle, the unrestrained vertical member lengths at the ends become longer; the load capacity decreases.

From Fig. 2, it is also apparent that the load capacities for K_F-values greater than 1.0 decrease. This can be attributed to the intersection points of the generatrices moving towards the top in the case of a smaller top ring diameter, therefore increasing the unrestrained vertical member lengths. When an additional unfavourable load is applied, the most efficient range for rotation angles remains around 105°, as the dashed lines show in the figure.

Hyperbolic structures: Shukhov's lattice towers – forerunners of modern lightweight construction, First Edition. Matthias Beckh.
© 2015 John Wiley & Sons, Ltd. Published 2015 by John Wiley & Sons, Ltd.

	Cylindrical shell $\varphi = 0°$	Hyperboloid $\varphi = 60°$	Hyperboloid $\varphi = 90°$	Hyperboloid $\varphi = 120°$
Perfect geometry	7.853	7.842	5.664	4.389
Imperfect geometry	6.226	7.640	5.549	4.342
Reduction	20.21 %	2.56 %	2.04 %	1.09 %

1

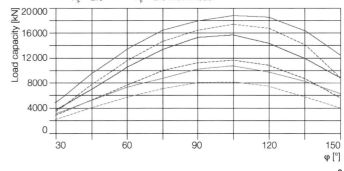

2

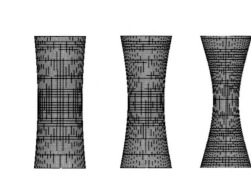

3

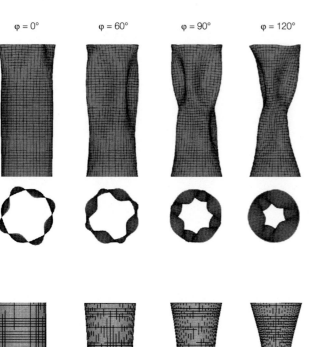

4

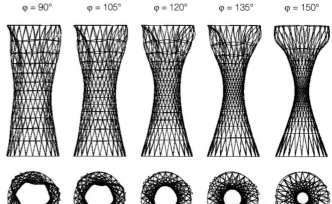

1   Comparison of the critical load factors for perfect and imperfect geometries
    of continuum shells of different curvatures (linear calculation)
2   Variant 1: Load capacities shown in relation to φ, linear calculation
3   Comparison of the buckling modes of different continuum shells
4   Investigated continuum shells (l. to r.): cylindrical shell, one-sheeted hyperboloid
    with rotation angles of 60°, 90° and 120°
5   Linear-linear calculated buckling modes of variant 1 in shown in relation to φ
    ($R_U = 5$ m; $K_F = 1.0$; n = 24)

5

Mesh variant 2: Construction used by Vladimir G. Shukhov

In contrast to the first mesh type, the horizontal intermediate rings in this Shukhov-built variant are positioned at equal vertical distances from one another and independently of the intersection points of the generatrices.

Influence of the rotation angle φ

As with the first mesh type, varying the rotation angle for a given number of intermediate rings and vertical members gives optimums in the range 90 to 120° for K_F-values of between 1.0 and 2.0. In the case of the equally spaced intermediate rings, the graphs are considerably flatter than before because the reduction of the unrestrained vertical member lengths means that the buckling of individual members is no longer determinant for very small or very large rotation angles. This rotation angle range still shows a clearly smaller load capacity than the range between 90 and 120° – evidence of the influence of the intersection points on the system stiffness.

The graphs of the second mesh type are clearly more discontinuous than those of the first variant. Because the rotation angle φ can increase independently of the central angle ψ, vertical member ends may not always meet in pairs at the top edge. Only when the rotation angle is a multiple of ψ, do triangular meshes occur at the top edge; otherwise trapezoidal meshes occur there, thus creating a more flexible edge. This is one reason for the sawtooth shape of

the linearly calculated graphs of the load capacity in Fig. 6a. An additional factor here is the superimposition of various other effects. Hence, for example, with a rotation angle φ of 105°; although the vertical member ends meet at the top edge, the unfavourable non-coincidence of the intermediate rings with the intersection points of the generatrices produces a smaller load capacity than with rotation angles of 90° or 120°.

Non-linear analyses

Building upon the two-stage linear calculation of load capacities, non-linear analyses were also carried out. If the load capacities are calculated in this way for a range of rotation angles, the optimum values also lie between 90 and 120°. The shape of the load capacity graphs is similar to that of the linear case, but they are clearly more continuous (Fig. 6b).

The influence of imperfections can be seen in the non-linear case. While with the linear analysis, the load capacity of the imperfect geometry compared with the perfect geometry drops to about 95%, the reduction of the load capacity with non-linear analysis is more severe. Fig. 6c shows the relationship between the ultimate loads calculated from the non-linear analyses of the perfect and imperfect structures. In general, the reduction is 20–30%.

Although according to the results from variant 1, it could be assumed that the load capacity for increasing K_F-values reduces, it is apparent from the linear analysis that the results for $K_F = 2.0$

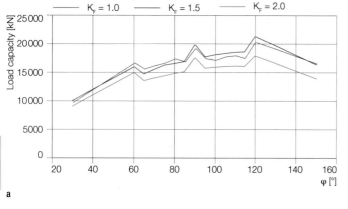

a

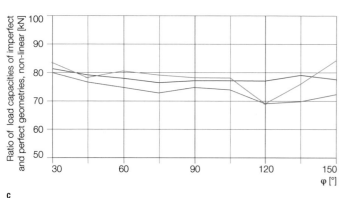

c

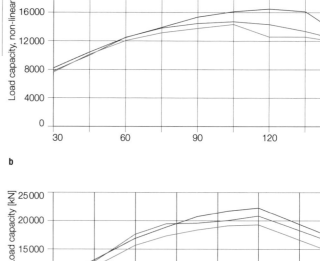

b

d

6

shows the least reduction, the load capacities for K_F = 1.0 and
K_F = 1.5 overlap and in some areas of the graph the values for
K_F = 1.5 are higher. The non-linear analysis shows this effect even
more clearly. One of the causes is that for larger K_F-values, that is to
say smaller top radii, the member inclination with reference to the
vertical decreases. The vertical member forces reduce for a given
vertical loading. On the other hand, the lower area is less
stiff because of the more widely spaced intersection points and
therefore the load capacities lower again at K_F = 2.0. Conclusive
statements are difficult to make, as behaviour greatly depends
on the particular configuration.

**Influence of the rotation angle φ with flexurally stiff intermediate ring
connections**
With the other variants investigated, described in the section
"Modelling and vertical member arrangement" (p. 46f.), the
connections of the vertical members at the intermediate rings –
as was usual with Shukhov – were modelled as pinned joints.
In comparison with those models, flexurally stiff connections
are now considered here.
The linear analysis gives rise to ultimate loads as shown in Fig. 6d;
the curves are generally more continuous than for the variants
with pinned connections discussed above. The ultimate loads are
from 6 to 16 % higher; on average there is a significant increase
of about 10 %.

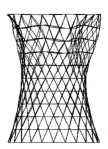

7

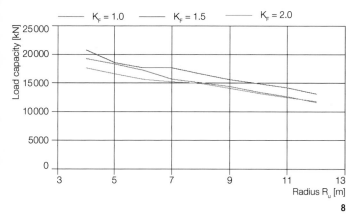

— K_F = 1.0　　— K_F = 1.5　　— K_F = 2.0

8

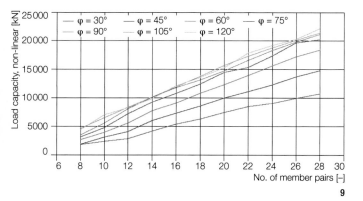

— φ = 30°　　— φ = 45°　　— φ = 60°　　— φ = 75°
— φ = 90°　　— φ = 105°　　— φ = 120°

9

6　Variant 2 (R_U = 5.0 m, IR = 10, n = 24)
　　a　Load capacities shown in relation to φ, linear calculation
　　b　Load capacities shown in relation to φ, non-linear calculation
　　c　Relationships between load capacities for perfect and imperfect geometries,
　　　　non-linear calculation
　　d　Load capacities shown in relation to φ, stiff intermediate ring connections,
　　　　linear calculation
7　Buckling modes for R_U = 3.0 m; 5.0 m and 9.0 m; K_F = 1.0; φ = 90°
8　Variant 2 (IR = 10; n = 24; φ = 90°):
　　Load capacities shown in relation to R_U for different K_F-values, linear
　　calculation
9　Variant 2: Load capacities shown in relation to the number of vertical member
　　pairs (R_U = 5.0 m; K_F = 1.0; IR = 10)

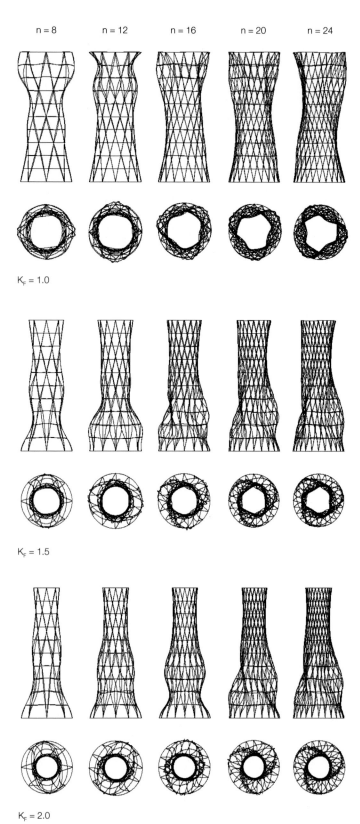

n = 8   n = 12   n = 16   n = 20   n = 24

$K_F = 1.0$

$K_F = 1.5$

$K_F = 2.0$

Influence of sizes of radii

For purely vertical loads, the load capacity decreases with increasing size of radius. This applies irrespective of the radii for all K_F-values. In the case of small radii, the buckling mode takes an oval or triangular shape at the affected heights. For larger radii on the other hand, the failure modes become increasingly short-waved as the system stiffness decreases. Fig. 7 shows examples of the buckling modes for three different radii with the same boundary conditions.

The decrease in the vertical load capacity for increasing radii size applies irrespective of the selected rotation angle φ (Fig. 8, p. 53).

Influence of the number of vertical member pairs n

The influences of the number of vertical members and the intermediate rings on load capacity were investigated for the mesh variant built by Shukhov. Fig. 9 (p. 53) shows the load capacity in relation to the number of vertical member pairs n. As expected, increasing the number of vertical members and consequently enlarging the overall cross-sectional area also increases the load capacity. In addition, the curves flatten out slightly for higher numbers of vertical members, because from a certain point the intermediate rings cannot accommodate further stabilising load. Fig. 10 shows the buckling modes for varying numbers of vertical members and a constant rotation angle.

To obtain a better expression of the efficiency of the structures, the ratio of the load capacity to the mass of steel is plotted on the ordinate in Fig. 13. After 20 vertical member pairs, no significant increase in the load capacity/mass ratio is apparent; the increase in self-weight continues at the same rate as the increase in load capacity.

Influence of the number of intermediate rings

An increasing number of intermediate rings produces a continuous increase in load capacity – in this case no flattening out of the curve is apparent (Fig. 14a). If the number of intermediate rings is plotted against the the load capacity/mass ratio as shown in Fig. 14b, the graphs show a continuous slight increase. An optimum value cannot be determined. As the rings are relatively light, they can produce a significant increase in the load capacity at comparatively little cost.

Influence of the stiffness of the intermediate rings

The stiffness of the intermediate rings has a huge impact on structural behaviour. In the analyses, the size of the angle profile of the intermediate rings increases in value from 10 to 300 % of the size of the vertical members (∑ 120/120/12 mm). The calculations show that increasing the intermediate ring profile size results in significant increases in load capacity. This applies generally for different rotation angles and K_F-values. With ten intermediate rings and $K_F = 1.0$, any increase in load capacity beyond a doubling of the size of the angle profile is not possible. From this point, the failure of the structure is not determined by global buckling (Fig. 11) but a unidirectional twisting of the vertical members. The stiff rings rotate with them in this case, as can be seen clearly in Fig. 12; the intermediate rings retain their shape however. The same also applies for $K_F = 2.0$. The failure point and with it the plateau area of the ultimate load is reached with a 1.4-fold increase in the leg length of the vertical member profile.

10

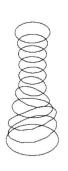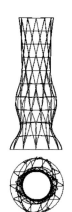

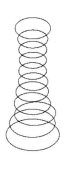

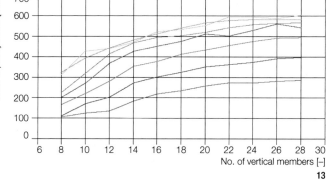

11

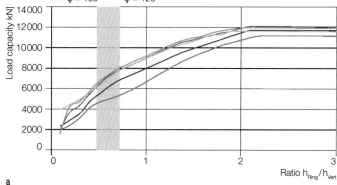

12

10 Buckling modes of variant 2 for various values of shape parameter K_F and number of vertical member pairs n in elevation and plan ($R_U = 5.0$ m; IR = 10; $\varphi = 30°$)
11 Buckling mode $K_F = 1.5$; $\varphi = 90°$; $h_{Ring} = 0.8\ h_{Vert}$
12 Buckling mode $K_F = 1.5$; $\varphi = 90°$; $h_{Ring} = 2.5\ h_{Vert}$
13 Variant 2 ($R_U = 5.0$ m; $K_F = 1.0$; IR = 10): Vertical member pairs to load capacity/mass ratio
14 Variant 2 ($R_U = 5.0$ m; $K_F = 1.0$; IR = 24):
  a Number of intermediate rings (IR) to load capacity
  b Number of IR to load capacity/mass ratio
15 Variant 2 ($R_U = 5.0$ m; IR = 10; n = 12):
  a Load capacities shown in relation to intermediate ring size, $K_F = 1.0$
  b Load capacities shown in relation to intermediate ring size, $K_F = 2.0$

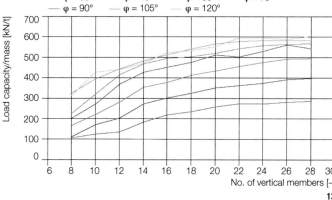

— $\varphi = 30°$ — $\varphi = 45°$ — $\varphi = 60°$ — $\varphi = 75°$
— $\varphi = 90°$ — $\varphi = 105°$ — $\varphi = 120°$

Load capacity/mass [kN/t]

No. of vertical members [–]

13

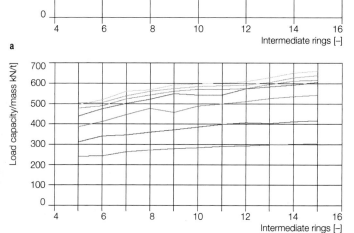

Load capacity kN

Intermediate rings [–]

a

Load capacity/mass kN/t

Intermediate rings [–]

b

14

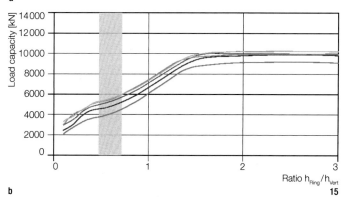

— $\varphi = 60°$ — $\varphi = 75°$ — $\varphi = 90°$
— $\varphi = 105°$ — $\varphi = 120°$

Load capacity kN]

Ratio h_{Ring}/h_{Vert}

a

Load capacity [kN]

Ratio h_{Ring}/h_{Vert}

b

15

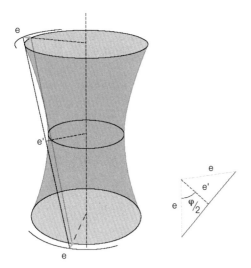

The areas shaded grey in the graphs show the range in which the intermediate rings used by Shukhov fall (Fig. 15). As a rule, their cross section height h_{Ring} is 50–75 % of that of the vertical members h_{Vert}. As shown in the section "Modelling and vertical member arrangement" (p. 46f.), standard angle profiles ($\sum 80/80/10$ mm) are used in the load capacity analysis of variant 2. Their size is two thirds of the vertical members – these are the profiles typically found in the structures. On examination, the graphs indicate that stiffer intermediate rings would produce a considerable increase in load capacity (Fig. 15, p. 55). Furthermore, the results make clear that the load level achievable through the use of stiff intermediate rings is almost independent of the rotation angle. With the intermediate rings normally used in variant 2, a clear relationship is shown to exist between load capacity and rotation angle, as was demonstrated earlier. The best results were obtained by creating as large a number of intersection points as possible, distributed as evenly as possible over the height. However, the role of the intersection points in determining the system stiffness becomes less important with increasing stiffness of the intermediate rings, which then mostly determine load capacities.

16

Influence of multilayer construction

As explained earlier, with the Shukhov-built variant the two vertical member families are arranged eccentrically in two planes. This creates a slight deviation from the theoretical geometry, which means that a generating straight line, which at the starting and finishing ends is displaced a distance e normal to the end rings, still has only an offset e' to the original position at the waist (Fig. 16). The maximum difference Δ between the theoretical geometry and the new vertical member position can be calculated as:

$$\Delta = e \, (1 - \cos(\varphi/2)) \tag{F01}$$

As the eccentricity between the two families with the Shukhov-built towers is constant, the vertical members do not exactly follow the parallel courses of the generatrices. As a result, the vertical members are slightly curved over their length, which leads to – albeit small – offset moments. The ultimate load is reduced with increasing vertical member eccentricity (Fig. 18).
The size of the vertical member eccentricity increases with the rotation angle. The graphs show the reduction in ultimate load for all rotation angles. However, the effect on the ultimate load is less for higher rotation angles because the structure – due to the greater number of intersection points – has a greater stiffness.

Influence of an additional horizontal load

If the structure is subjected to a combination of vertical load and a horizontal component of 5 % at the top ring, the mode of failure depends greatly on the geometry. While the previously undertaken parametric studies found that buckling failure almost always determined the load capacity under purely vertical loading, reaching the yield stress is often decisive with the above load combination. If buckling is determinant then its effects are no longer distributed over a large area of the circumference of the structure but occur locally on the compression side (Fig. 17). Stress failures mainly arise in the case of small radii and are a result of the smaller internal

$K_F = 1.0$ $K_F = 1.5$ $K_F = 2.0$

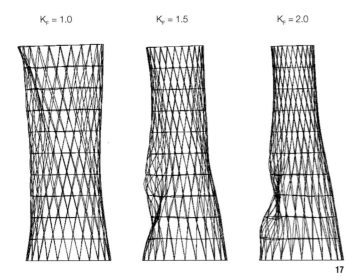

17

lever arm of the structure. While with otherwise identical boundary conditions, primary buckling failure is relevant in the case of $R_U = 5.0$ m; at a radius of $R_U = 3.0$ m the influence of stress failures can be clearly recognised (Fig. 19 a and b). Moreover, the number of intermediate rings is also relevant to the mode of failure. As the number of rings increases, the structure stiffens in such a way that the vertical members are loaded to their yield stress and stress failure plays more of a role in determining the load capacity. As with all the structures under purely vertical loading investigated earlier, the range of most favourable rotation angle for all K_F-values lies between 90 and 120°.

Summary of the results from variant 2

In summary for variant 2, the Shukhov design with intermediate rings at constant vertical spacings shows the following dependencies:

- The most favourable range of rotation angle φ for all K_F-values lies between 90 and 120°. This applies equally to vertical as well as combined vertical-horizontal load actions.
- For the cross sections forming the basis for the investigations, a rotation angle range that provides as great a number of intersection points as possible and distributes them as evenly as possible over the height of the structure is particularly favourable.
- An alternative flexurally stiff design for the intermediate ring connections to the vertical members increases the load capacity by about 10 %.
- In contrast to variant 1, the load capacities for K_F-values of 1.5 are generally higher than those for $K_F = 1.0$, because of the reduced inclination of the vertical members to the vertical.
- Increasing the number of vertical members increases the load capacity. Although under the assumed boundary conditions, the load capacity does not increase beyond an amount of about 16 vertical member pairs.
- Increasing the number of intermediate rings produces a continuous increase in the load capacity as well as a steady increase in the load capacity/mass ratio.
- Enlarging the cross sections of the intermediate rings can increase load capacity significantly. The load capacity is more than double the structure initially investigated.
- With greater intermediate ring stiffness, the influence on the load capacity of the number of intersection points, which is otherwise determinant, diminishes. The achievable load capacities converge on a value irrespective of the rotation angle.
- A slightly disadvantageous influence of the slightly curved deviation of the vertical members from the straight caused by the multilayer construction can be witnessed.

Mesh variant 3: Discretisation as reticulated shells

With this mesh variant, the vertical members do not follow the directrix of the hyperboloid but are arranged so that they form a homogeneous mesh of triangles with intermediate rings. At the base of the structure, the meshes are equilateral triangles with a height that sets the vertical distance between the intermediate rings. This vertical distance remains constant over the height of the structure. As a result the meshes above this level are isosceles triangles and

16 Eccentricity arising from multilayer construction
17 Variant 2: Modes of failure for additionally applied horizontal load in the negative x-direction ($R_U = 5.0$ m; IR = 10; n = 24; φ = 75°)
18 Variant 2: Load capacities shown in relation to the vertical member eccentricity ($R_U = 5.0$ m; $K_F = 1.0$; IR = 5; n = 24)
19 Variant 2: Load capacities shown in relation to φ with 5 % horizontal load
a $R_U = 3.0$ m; IR = 10; n = 24
b $R_U = 5.0$ m; IR = 10; n = 24

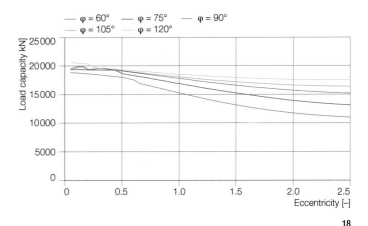

18

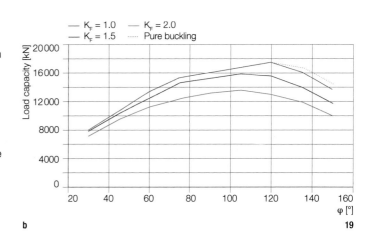

a

b

19

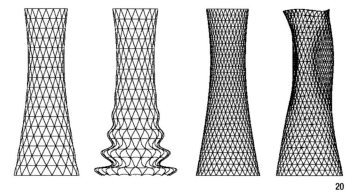

20

become more acute as they approach the waist. In contrast to the previous mesh variants, the vertical members here are in the same layer but are joined coaxially at the node points.

Influence of the number of vertical members on the load capacity:

The advantage of this third type of mesh lies in the slight changes in direction of the vertical members at each node point. The changes in direction allow the horizontal actions applied to the lattice mesh to be distributed primarily by normal forces. This means the structural behaviour of this form of mesh is the closest to that of a doubly curved reticulated shell. Because of the form of mesh, this type of structure is designed exclusively with pinned members as long as the top ring is flexurally stiff. Variants 1 and 2 would be kinematic in this case. This geometry requires a butt joint for each vertical member at every node, which significantly increases assembly costs.

As Fig. 20 shows, the system fails in the case of a low number of vertical members by individual member buckling. With a greater number of vertical members and consequently narrower mesh width, however, the structure fails over a larger area by global buckling. The buckling mode is then comparable with that of an equivalent continuum shell. In Fig. 21a, the transition from individual member failure to global failure has a pronounced kink, which occurs in the selected example at 28 vertical member pairs. From this point, the global buckling mode is for the first time weaker than the corresponding individual member buckling modes. The associated load capacity/mass ratio shows that from this point no further increase in the efficiency of the structure is possible (Fig. 21b).

Influence of the node stiffness on the load capacity

The stiffness of the connections is especially important as the vertical members in this variant all lie in one plane and are connected at nodes. In the earlier analyses, the node points were assumed to be flexurally stiff. As in practice these have a certain amount of flexibility, the question arises of how node stiffnesses should be correctly modelled.

As already mentioned, variant 3 could in theory be designed out of pure truss members supported only on their longitudinal axes. The other extreme is a 100% flexurally stiff restraint of the vertical members at the nodes. The real situation is somewhere between these limits. The reduced stiffness of the nodes can be modelled using torsional springs, but this process is very complex for three-dimensional structures. One alternative method works by reducing the vertical member cross section in the area of the node. [4] In a similar way to this method, the vertical members of variant 3 are now each subdivided into six sections and the end elements allocated a variable radius. While the full vertical member cross section represents a flexurally stiff member end, small cross section radii can model almost-pinned connections.

Using the equivalent system of a beam fixed at both ends (Fig. 24) under a uniformly distributed load, the fixed-end moments and the node stiffnesses resulting from them are now calculated for the reduced radii (Fig. 25).

The modes of failure were calculated for the example of a tower structure with $R_U = 5.0$ m, $K_F = 1.5$, $\varphi = 50°$ and 32 vertical member pairs. The maximum vertical member length (for variant 3 this is the

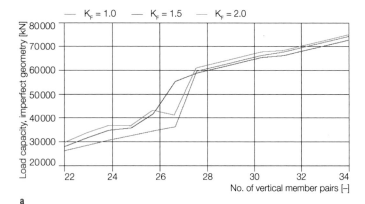

a

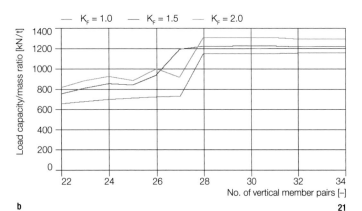

b

21

maximum distance between nodes) in this case is about 1 m.
For the buckling modes shown in Fig. 22, it is apparent that
the number of buckles increases with decreasing node stiffness;
a behaviour which can also be observed with continuum shells
as their thickness decreases.
Fig. 23 shows the load capacities in relation to relative node stiffness.
After the node stiffnesses have been determined from a model test,
the reduction in load capacity of the flexurally stiff structure can be
estimated for a future tower built in the form of variant 3.

Summary and comparison of the results

The surface of a one-sheeted hyperboloid was divided into three
different meshes to investigate the relationships and interactions
between the form and the structural behaviour of hyperbolic
lattice structures. The intermediate rings of variant 1 are positioned
at the intersection points of the vertical members; with variant 2,
which Vladimir Shukhov used for his tower structures, the intermediate rings are positioned at equal vertical intervals over the
height of the structure. Variant 3 is different to the others in the
following respects: Here the intermediate rings are spaced at equal
vertical intervals and the node points placed around adjacent
intermediate ring levels such that a uniform isosceles mesh
is created (Fig. 19, p. 44).
Comparing the results of the ultimate loads for variants 1 and 2, it can
be seen that Shukhov achieves clearly higher ultimate loads through
the evenly spaced arrangement of the intermediate rings. This effect is
particularly significant for higher K_F-values. The range of the
structurally most favourable rotation angle φ is between 90 and
120° for both variants, and applies for vertical loads and for vertical
and horizontal load combinations.
Assessing the economy of the investigated systems can be done
by comparing the load capacity/mass ratios. Generally applicable
conclusions are difficult to draw due to the vast number of influential
parameters, but some basic tendencies can be recognised. The
achievable load capacity/mass ratio of mesh type 2 was taken as the
basis for an example calculation. The first mesh type achieved 77 %,
the third 170 % of this value. If node stiffness is taken into
consideration, the advantages of the third variant become less
absolute; from a relative node stiffness of 70 %, the material cost is
again comparable with that of variants 2 and 3.
The question of which makes the most efficient use of material
depends primarily on the task and nature of the structure: with very

20 Buckling modes of the third mesh type with 18 and 32 vertical member pairs
   (R_U = 5.0 m; K_F = 1.5; φ = 90°)
21 Variant 3 (R_U = 5.0 m; φ = 60°)
   a Load capacities shown in relation to the number of vertical member pairs
   b Load capacity/mass ratio shown in relation to the number of vertical
   members
22 Buckling modes at 100 % (l.) and 43 % (r.) Node stiffness
23 Variant 3: Non-linear calculated load capacity shown in relation to
   node stiffness
24 Model for calculating the reduced edge stiffness
25 Fixed-end moments and relative node stiffnesses shown in relation
   to radius for vertical member length l = 0.995 m (R_U = 5.0 m; K_F = 1.5;
   n = 32; φ = 50°)

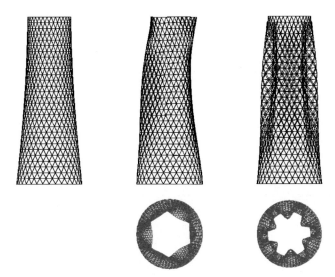

22

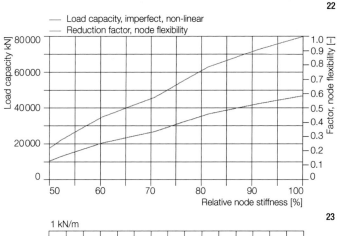

23

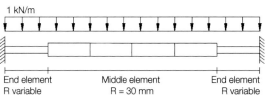

End element   Middle element   End element
R variable   R = 30 mm   R variable

24

Factor, radius [–]	Radius [mm]	I_P [cm⁴]	Fixed-end moment [kNm]	I_P [%]	Rel. node stiffness [%]
0.10	3.00	0.013	0.34	0.01	42.874
0.20	6.00	0.204	0.35	0.16	43.131
0.30	9.00	1.031	0.35	0.81	44.224
0.40	12.00	3.258	0.38	2.56	47.032
0.50	15.00	7.953	0.42	6.25	52.381
0.60	18.00	16.490	0.48	12.96	60.501
0.70	21.00	30.553	0.57	24.01	70.664
0.80	24.00	52.122	0.65	40.96	81.455
0.90	27.00	83.489	0.73	65.61	91.501
1.00	30.00	127.251	0.80	100.00	100.000

25

high loads and a correspondingly high number of vertical members, the third variant is the best. If the determinant load is small and therefore the number of vertical members is also small, then the coarseness of the mesh means that variant 3 structures do not fail by shell buckling. Instead they fail by individual member buckling; their load capacity/mass ratio is comparable with the second variant. However, the continuous vertical members and the simpler design of the node points reduce the complexity of fabrication of the second variant. Therefore, given the same material costs, the second variant is preferable to the third.

Structural analysis of selected towers built by Vladimir G. Shukhov

The next section contains examples of analyses and calculations of selected Shukhov water towers. The calculations are two-stage linear/linear and non-linear/non-linear in each case. The structures were selected to provide the longest spread of years of construction and the greatest diversity of form.

The self-weight of the load-bearing structure was taken from the values calculated by the software Ansys and the known weights recorded in the papers and design documentation of the water towers. The totals obtained were increased by 10 % to take into account the weight of items not directly included (e.g. rivets, joint plates, connections, pipes etc.). The weight of the water is then added to provide the total vertical loading. An equivalent horizontal load of 5 % of the vertical load is applied to the top ring to model the horizontal wind load. This simplified assumption was applied in comparative calculations for different tank sizes and is a conservative value for the effective total wind load acting on the tower and tank. No partial safety factors are applied to the loading side; the level of safety under the given loading is then calculated. In the absence of more accurate knowledge about the steel used, the calculations assumed grade S 235 – certainly an optimistic assumption for the early towers. As before, a partial safety factor of 1.1 was applied to the material side; the applied imperfections were 50 mm.

Water tower for the All-Russia Exhibition in Nizhny Novgorod (1896)

As the first completed hyperbolic lattice tower, the water tower for the All-Russia Exhibition in Nizhny Novgorod is of special significance. In spite of the comparatively small tank exerting very little load on the structure, it has more vertical members than any other built tower. As a consequence the number of intersection points of the vertical members is particularly high and the lattice mesh correspondingly fine. Fig. 26 c summarises the geometry, cross sections and loading of the tower.

Fig. 26 a shows the member model used for the calculations, Fig. 26 b the non-linear calculated buckling modes for both load cases. As can be seen without difficulty, buckling occurs in the bottom half of the lattice surface; the top part is very stiff due to the fine mesh pattern (Fig. 12, p. 39). When near the point of stability failure, the tower cross section ovalises at certain heights.

For vertical loading, non-linear calculations give an ultimate load of 5034 kN, for a combination of vertical and horizontal loads, the ultimate load is 3957 kN. With a purely vertical load of 1553 kN, the safety factor is calculated as 3.24 and as 2.54 for the combination of vertical and horizontal loads (Fig. 26 d).

Water tower in Mykolaiv (1907)

The water tower in Mykolaiv has by some margin the largest tank volume of all the Shukhov towers constructed to date and at the time was the largest Intze water tank in Russia. The original structural calculations are examined in the section "Design and analysis of Shukhov towers" (p. 66ff.), while a summary of the key data can be found in Fig. 27 c (p. 62).

Fig. 27 a (p. 62) shows the member model used for the calculations; Fig. 27 b (p. 62) shows the non-linear calculated buckling modes for both load cases. Under purely vertical loading, the intermediate rings in the lower part of the tower distort into a slightly triangular shape. For vertical loading, non-linear calculations give an ultimate load of 11,703 kN, while for a combination of vertical and horizontal loads, the ultimate load is 9719 kN. With a purely vertical load of 6980 kN, the safety factor is calculated as 1.70 and as 1.39 for the combination of vertical and horizontal loads (Fig. 27 d, p. 62).

Water tower in Tyumen (1908)

The water tower in Tyumen, which is discussed in the chapter "Design and analysis of Shukhov towers" (p. 66ff.), was considered a particularly light and cost-efficient structure in its day. In a book about water towers in Russia, this tower is identified as having the smallest "lightness coefficient", which is defined there as the ratio between total weight and tank volume. [5]

Fig. 28 a (p. 63) shows the member model used for the calculations; Fig. 28 b (p. 63) shows the non-linear calculated buckling modes for both load cases. For vertical loading, non-linear calculations give an ultimate load of 6205 kN, while for a combination of vertical and horizontal loads, the ultimate load is 5288 kN. With a purely vertical load of 4413 kN, the safety factor is calculated as 1.40 and as 1.20 for the combination of vertical and horizontal loads (Fig. 28 d, p. 63).

Water tower in Dnipropetrovsk (1930)

The water tower in Dnipropetrovsk collapsed in 1930, one week after it was erected. Based on the review by the Ukrainian engineer V. A. Djadjuša [6] discussed in the section "The collapse of the water tower of Dnipropetrovsk" (p. 84f.), which blamed the pinned design of the connections between the verticals and the intermediate ring, a variant with flexurally stiff ring connections was investigated in addition to the built structure.

For vertical loading, non-linear calculations give an ultimate load of 3111 kN, while for a combination of vertical and horizontal loads, the ultimate load is 2482 kN. With a purely vertical load of 3160 kN, the safety factor is calculated as 0.98 and as 0.79 for the combination of vertical and horizontal loads (Fig. 29 d, p. 64). The vertical load acting alone was enough to cause a stability failure of the structure.

---

26 Water tower, Nizhny Novgorod (RUS) 1896
    a Finite-element models
    b Buckling shapes, non-linear, vertical load (l.) and vertical and horizontal load (r.)
    c Summary of geometry, cross sections and loads
    d Load-displacement diagram (L-D diagram)

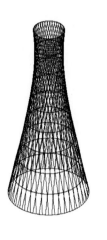

a

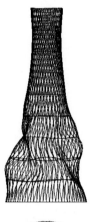

b

Nizhny Novgorod (1896)

Geometry		Cross sections		Loading	
Height	25.60 m	Vertical rods	L 75/75/10 mm L 50/50/8 mm	Weight of structure	291.3 kN
Number of vertical members/rings	80/8	Ring	L 75/75/10 mm	Weight of tank	91.7 kN
Diameter of bottom	11.07 m			Tank volume	117 m³ (1170 kN)
Diameter of top	4.27 m				
Rotation angle	94.5°			**Total vertical load**	**1553 kN**

c

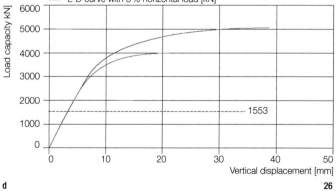

d

26

Variant with flexurally stiff ring connections

For vertical loading, non-linear calculations give an ultimate load of 4151 kN, while for a combination of vertical and horizontal loads, the ultimate load is 3323 kN. The corresponding buckling modes are shown in Fig. 30, p. 65. With a purely vertical load of 3160 kN, the safety factor is calculated as 1.31 and as 1.05 for the combination of vertical and horizontal loads (Fig. 31, p. 65).

Comparison of the results of the investigated water towers

The load capacities of the four investigated towers for vertical and vertical-horizontal loads are summarised in Fig. 32 (p. 65). The results lead to the conclusion that the achieved safety factors reduced over time with each successive design.

The first water tower in Nizhny Novgorod had a calculated safety factor of 2.54 for the vertical and horizontal load combinations and therefore had a safety factor that would have satisfied today's requirements. The safety factors of the Shukhov towers built in earlier years are also around this value, for example the water towers in Lysychansk (1896) and Yefremov (1902; Fig. 34, p. 91), which are not dealt with in further detail here, were calculated as having safety factors greater than 3.00 [7].

On the other hand, the achieved safety factors of the towers in Mykolaiv and Tyumen for the load case combinations at 1.39 and 1.20 are rather low. The two structures with their large tank volumes are a clear advance on the earlier installed water capacities – in the case of Mykolaiv, this was a five-fold increase in the top load of earlier structures. Obviously the confidence in the design and construction after more than ten years' experience allowed the limits to be pushed further and the safety reserves of the system to be continuously reduced. To this can be added the improvement in the available steel quality and permissible steel stresses, which are discussed in the chapter "NiGRES tower on the Oka" (p. 96ff.). For example, the permissible steel stresses used in the historical calculations for the water tower in Tyumen are around 25 % higher than those used for the Mykolaiv tower.

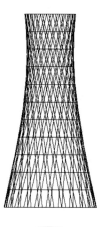
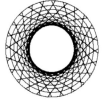

a

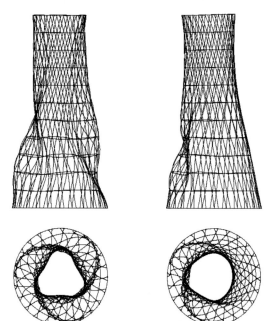

b

Mykolaiv (1907)

Geometry		Cross sections		Loading	
Height	25.6 m	Vertical rods	L 120/120/12 mm L 100/100/10 mm	Weight of structure	420.0 kN
Number of vertical members/ rings	48/10	Ring	L 80/80/10 mm	Weight of tank	410.0 kN
Diameter of bottom	12.08 m			Tank volume	615 m³ (6150 kN)
Diameter of top	7.01 m				
Rotation angle	82.5°			**Total vertical load**	**6980 kN**

c

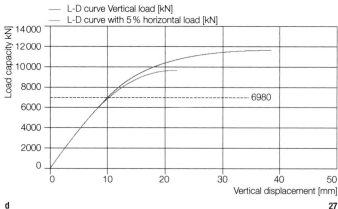

d

27

The water tower in Dnipropetrovsk does not achieve a safety factor greater than 1 for purely vertical loading. This tallies with the story of its collapse, which can be attributed to stability failures and was discussed in contemporary specialist literature (see "The collapse of the water tower in Dnipropetrovsk", p. 84f.). Using flexurally stiff connections to the intermediate ring, the load capacity would be increased by around 30 % – even though by doing this an acceptable level of safety would still not be achieved. This could have been done without great cost by the use of stronger intermediate rings, as the calculations in the section "Influence of the stiffness of the intermediate rings" (p. 54ff.) show. However, the choice of intermediate rings was not based on calculations. In the main, the depth of cross section of the intermediate rings was between 50 and 75 % of the depth of the vertical member cross section. In the case of large diameter intermediate rings – as is the case here – this relationship, which is determined independently of the radii, leads to cross sections with too little flexural stiffness.

27 Water tower, Mykolaiv (UA) 1907
  a Finite-element models
  b Buckling modes, non-linear, vertical load (l.) and vertical and horizontal load (r.)
  c Summary of geometry, cross sections and loads
  d Load-displacement diagram
28 Water tower, Tyumen (RUS) 1908
  a Finite-element models
  b Buckling modes, non-linear, vertical load (l.) and vertical and horizontal load (r.)
  c Summary of geometry, cross sections and loads
  d Load-displacement diagram

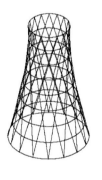
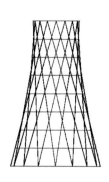

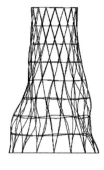

a

b

Tyumen (1908)

Geometry		Cross sections		Loading	
Height	19.2 m	Vertical rods	L 110/110/12 mm L 100/100/12 mm	Weight of structure	169.0 kN
Number of vertical members / rings	32/7	Ring	L 75/75/10 mm	Weight of tank	144.0 kN
Diameter of bottom	12.19 m			Tank volume	410 m³ (4100 kN)
Diameter of top	6.25 m				
Rotation angle	67.5°			**Total vertical load**	**4413 kN**

c

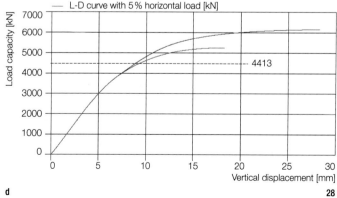

— L-D curve Vertical load [kN]
— L-D curve with 5 % horizontal load [kN]

d

28

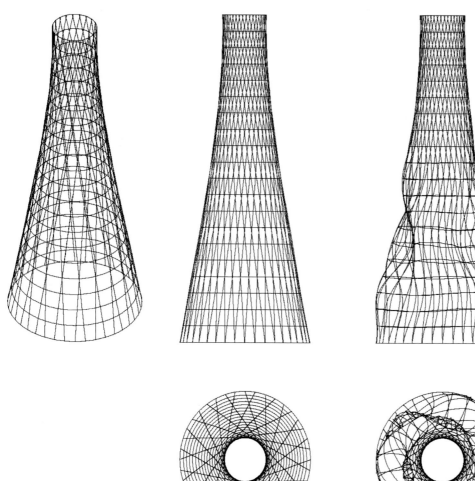

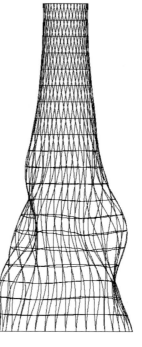

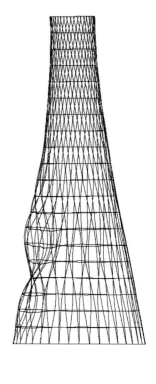

a

b

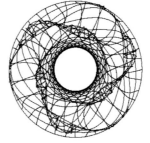

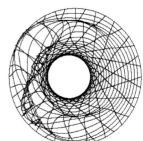

Dnipropetrovsk (1930)

Geometry		Cross sections		Loading	
Height	43.5 m	Vertical rods	L 100/100/12 mm L 100/100/10 mm	Weight of structure	457.6 kN
Number of vertical members / rings	40/19	Ring	L 75/75/10 mm	Weight of tank	242.4 KN
Diameter of bottom	18.0 m			Tank volume	246 m³ (2460 kN)
Diameter of top	6.0 m				
Rotation angle	78°			**Total vertical load**	**3160 kN**

c

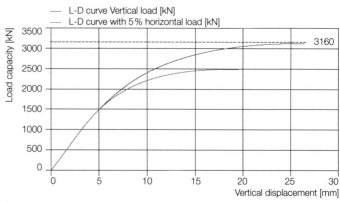

d

29

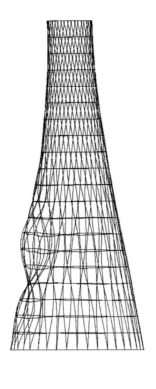

Even if simplifications have been made for the analysed examples, and the selection may not be representative, the results show that the structures built later (for the loads considered) have a very low level of safety. This is all the more valid when it is considered that today's steel grade S 235 was used for the calculations – which is probably rather optimistic for the earlier examples from the period studied.

29 Water tower, Dnipropetrovsk (UA) 1930
    a Finite-element models
    b Buckling modes, non-linear, vertical load (l.) and vertical and horizontal
       load (r.)
    c Summary of geometry, cross sections and loads
    d Load-displacement diagram (as built)
30 Buckling modes (variant with flexurally stiff ring connections), non-linear, vertical load (l.) and vertical and horizontal load (r.)
31 Load-displacement diagram, water tower in Dnipropetrovsk (flexurally stiff ring connections)
32 Summary of the load capacities and resulting safety factors

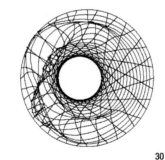

30

31

	Constructed [–]	Tot. Vert. load [kN]	Vert. load cap. [kN]	Safety factor [–]	Vert.-horiz. load cap. (horiz. = 5 %) [kN]	Safety factor [–]
Nizhny Novgorod	1896	1553	5034	3.24	3957 (198)	2.54
Mykolaiv	1907	6980	11703	1.70	9719 (486)	1.39
Tyumen	1908	4413	6205	1.40	5288 (264)	1.20
Dnipropetrovsk	1930	3160	3111	0.98	2482 (124)	0.79
With stiff rings		3160	4151	1.31	3323 (166)	1.05

32

Design and analysis of Shukhov's towers

To gain a deeper insight into Shukhov's methods of design and the underlying theory, the calculations for his hyperbolic lattice towers and their static models are analysed and evaluated in the following sections. The basic information was obtained from the archives of the Russian Academy of Sciences in Moscow, the city archives at Nizhny Novgorod, the private archives of Elena Shukhov and the Stuttgart University library catalogue.

The design process for his water towers is also reconstructed, the development of the tower design over time investigated and some details of construction briefly explained. To obtain a better understanding and set the historic context, the following section presents a general picture of the development of steel water towers and the state of structural engineering knowledge at the time of Shukhov.

The development of steel water tanks and water towers

It is helpful in establishing the historical context and the significance of Shukhov's water towers to examine the development of steel water tanks, because the different forms of construction of these containers always have their own characteristic requirements and consequences on the supporting structure. Extensive consideration of the architectural history and technical development of these structures can be found, for example, in contemporary papers on steel construction (various authors including Eduard Kottenmeier [1]) and in the publication "Ursachen und technische Voraussetzungen für die Entwicklung der Wasserhochbehälter" [2].

The development of water towers, which have their predecessors in the hydraulic structures of the Middle Ages and early water supply systems for towns and villages, started in the middle of the 19th century. The same era also saw the construction of mostly cast iron flat-bottomed tanks, which often took the rectangular box shape of their wooden predecessors (Fig. 2a). One of the first cylindrical flat-bottomed tanks in Germany to be built completely out of wrought iron was the 1868 water tower in Halle an der Saale (464 m³ capacity, Fig. 2b).

This tank is made out of singly curved plates riveted together at their overlapping edges, which are then sealed with spun yarn and red lead. The crucial disadvantage of the flat-bottomed tank is its supporting grillage beam system, which is loaded in bending, therefore material-intensive and costly, and also makes maintenance of the underside of the tank difficult. In 1855, Frenchman Jules Dupuit was the first to design the tank bottom as a suspended calotte and therefore make the cylinder walls and the bottom into predominantly tensile members. The first water tanks with a suspended bottom were built from the end of the 1860s for French railway companies and had medium storage capacities. One example is the elevated tank with a 150 m³ capacity built for the Midi-Ouest railway (Fig. 2c). [3] The bottom plates are typically fastened with an angle profile (120°) to the cylindrical tank wall and connected by another angle profile – in this case cast iron – to a support ring, which distributes the vertical forces into the substructure (Fig. 4).

The shallow inclination of the calotte bottom places the support ring for the suspended tanks under considerable compressive stress. In tanks with larger volumes, the rise and fall of the water level causes the support ring to deflect, exhibiting a wide range of movement, which generally leads to the tanks leaking. [4] Over the years, engineers sought by various design measures to eliminate this construction problem without ever finding a fully satisfactory solution.

A substantial improvement was brought about only by a later invention patented by the Aachen engineer Otto Intze. For a water tower built in 1883 in Remscheid, he incorporated a conical area into the tank cross section from the bottom edge of the cylinder wall to an inner support ring. By inverting the then conventional suspended bottom to form a supporting bottom, he creates a compression-loaded domed calotte (with stiffening ribs): in geometrical terms a spherical cap, with the horizontal bearing force components acting outwards on the support ring. The horizontal forces are cancelled out if the angles of the conical section are appropriately chosen to suit the tangent angle of the domed calotte. In this case, the support ring has to be designed only for the vertical loads and therefore the task is much simpler. For tank volumes of over

Hyperbolic structures: Shukhov's lattice towers – forerunners of modern lightweight construction, First Edition. Matthias Beckh.
© 2015 John Wiley & Sons, Ltd. Published 2015 by John Wiley & Sons, Ltd.

1 Water tower, Lugovaya near Moscow (RUS), condition in 2008
2 Elevated water tanks
  a Maisons-Lafitte near Paris (F) 1850
  b Halle an der Saale (D) 1868
  c For the French Midi-Ouest railway (F) ca. 1865
3 Horizontal forces acting on the support ring for suspended bottom tanks
4 Different methods of fastening the suspended bottom to the support ring
5 Cancelling out of the horizontal force component with Intze tanks (type I)

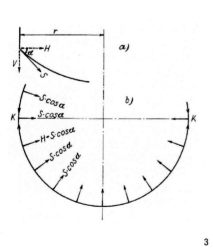

a

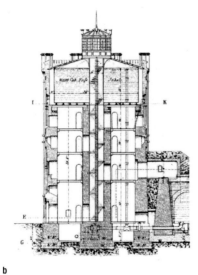

b

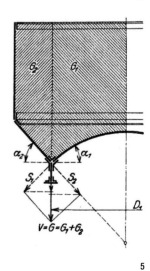

c

2

3

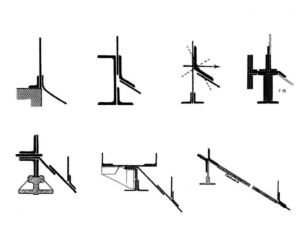

4

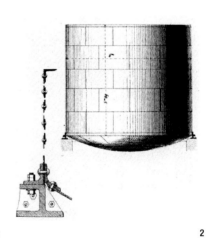

5

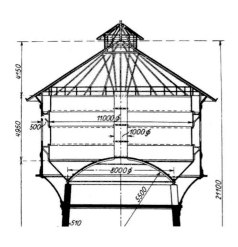
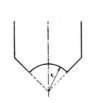

a

b

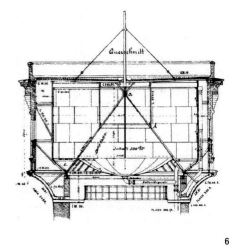

6

6 Water tanks to the Otto Intze design: type I with a supporting bottom (a) and type II incorporating a suspended bottom (b)
7 Water tower, Paris (Illinois, USA) 1897
8 Barkhausen tank, Dortmund (D) 1899
9 Reinforced concrete water tank by Eduard Züblin, Scafati (I) 1897
10 Reinforced concrete water tower, Singen (D) 1907

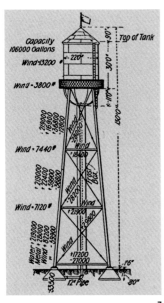

7

8

9

10

1000 m³, this leads to material savings of around 10 % and for smaller capacity tanks around 25 %. [5] Intze improved his tank shape in 1885 by replacing the supporting bottom with a second upstand to his cone in combination with a suspended bottom (Intze type II, Fig. 6 b). This did away with the large construction depth of the supporting bottom and resulted in a considerable increase in tank capacity. Moreover, it dispensed with the need to fabricate the complex buckling stiffeners required for the supporting bottom. The new form of construction proved very successful for the Aachen company F. A. Neumann, who acquired the patents from Intze. Between 1891 and 1902, the company manufactured 85 series-produced railway water towers to the Intze type I design alone, with standardised tank sizes of between 15 and 1000 m³. [6]

The Intze tanks continued to be the dominant construction type for water tanks in Germany until the start of the 20th century. In 1898, Georg Barkhausen developed a new type of tank, which was a further development of the earlier suspended bottom type. It became known as the Barkhausen tank and consists of a bottom hemisphere which continues into the cylindrical section without an abrupt change in direction and therefore required no compression ring. The first water tower with a Barkhausen tank was built in 1899 for the Minister Stein mine [7] in Dortmund (520 m³, Fig. 8). In contrast to the earlier iron water tanks, which are generally supported by a brickwork structure, the supporting structure here was made of iron. The columns connect to the cylindrical part of the tank and use the supporting action of the walls. This type of solution cannot be applied to the Intze tanks because the inner support ring cannot provide direct support to the tank walls and therefore an additional heavy ring beam would have been needed.

Similar structures to the Barkhausen tank with inclined supporting structures were designed in the USA by the Chicago Bridge and Iron Company [8] (water tower in Paris, Illinois, 1897; Fig. 7).
A further development of the sealed iron tank is the Barkhausen tank with a top hemisphere. These were built from 1906 by the Dortmund company A. Klönne, the licence holder of the Barkhausen tank. This was shortly followed by the spherical tank (Klönne type), which minimised the large pressure differences of the Barkhausen tank and by virtue of the resistance to buckling of its doubly curved plates also required no stiffening ribs.

The first reinforced concrete water towers appeared from the end of the 19th century; the earliest examples being by reinforced-concrete pioneers Joseph Monier (in Alençon, 1873, 180 m³) and François Hennebique (in Dinard, 1895, 500 m³). [9] Engineers such as Eduard Züblin took them further and developed sophisticated and successful structures (e. g. in Scafati, 1897, 80 m³; Fig. 9). During the following years, this new, more cost-effective material quickly became popular for the construction of water towers and increasingly threatened the predominance of iron. In contrast to Germany and western Europe, where the development of iron water towers stood still from around 1910 and sharply declined in significance, the Intze, Barkhausen and spherical tank designs were revisited in the USA and further developed into voluminous steel tank structures such as the Horton and the obloidal tank.

The water towers of Vladimir G. Shukhov

In addition to the fields of application mentioned earlier, Shukhov used his hyperbolic lattice structures mainly for water towers. The Russian construction engineering company Bari relied on systems developed and proven in Germany and western Europe for its water towers. Cylindrical flat-bottomed tanks, some suspended bottom tanks, were used as well as Intze type I (used for the first tower in Nizhny Novgorod) and type II tanks. The Givartovsky water tower in Moscow with a 3 foot rise and 18 foot span is a suspended bottom tank.

Shukhov's water towers are probably the first tower structures built entirely out of steel to be used for this task. When the tower in Nizhny Novgorod was built in 1896 for the All-Russia Exhibition, tanks in western Europe and the USA relied mainly on masonry construction for their supporting structure. The previously mentioned examples of steel-framed construction used for the water towers in Dortmund and Paris (Illinois, USA) followed a few years later. However, compared with these more conventional structures, the Shukhov design combines a number of advantages:

• The concise form, clear detailing and slender cross sections lead to an architecturally pleasing structure. This applies not only in comparison to conventional frame structures but also to the masonry structures normally encountered in western Europe, which some people found ugly, jokingly saying their shape looked as if they had from "water on the brain". [10] Shukhov towers were often used to provide a element of urban design in Russia (see "Water towers", p. 21).
• The continuous support provided to the top ring, which carries the tank, allows the lattice structure below to be much more slender and efficient in its use of materials than conventional framework structures.
• Shukhov's design is ideal in particular for water towers with modern Intze tanks because they require a continuous support. Point supports from a conventional framework tower would have required a considerably heavier ring beam and therefore cancelled out the advantages of the Intze tank. For this reason, steel framework structures are used in western Europe and the USA primarily for suspended-bottom and Barkhausen tanks, because in these systems the top vertical support members connect to the tank walls and therefore an additional compression ring can be omitted.
• Their small cross section sizes means the profiles are easy to transport and quick to erect.
• The efficiency of the structures is much higher than other systems, as investigations at the time prove.

Particularly interesting in this context is a table in Dimitrij Petrov's book [11] which investigates the costs in relation to tank volume and height of selected examples. The parameter used for the comparison is a type of efficiency factor: $m = P/(h \cdot V)$, where P is the total cost of the tower structure, h the tower height and V the tank volume. In his comparison of the various tower types, Petrov arrives at the following values:
• Stone and brick construction   m = 0.075 – 0.234
• Conventional iron construction   m = 0.098 – 0.133
• Towers to a Shukhov design   m = 0.053 – 0.069

Surprisingly, one tower stands out from the others in the study in terms of its efficiency: the water tower for the Maggi factory in Singen, which was built to height of 48 m using reinforced concrete in 1907 (Fig. 10, p. 68). It has an m factor of only 0.048, which is an indication of why reinforced-concrete water towers became increasingly popular in western Europe in the 1910s.

Development of structural analysis and engineering design methods in the 19th century

So that Shukhov's methods of structural engineering design discussed later in this chapter can be better set in context and evaluated, a brief summary of the state of structural analysis at the end of the 19th century is given below. The summary draws in particular on the extensive literature on this subject by Bill Addis, Karl-Eugen Kurrer and Hans Straub. [12]

The many individual discoveries in the 18th and up to the beginning of the 19th century in the field of structural analysis were brought together to form a separate discipline above all in the book "Mechanik der Baukunst" by Claude Louis Marie Henri Navier, which was published in 1826 and contains the theory of bending. Hans Straub wrote about this publication: "Its pioneering value lies not only in the numerous new methods that the work contains, but perhaps much more in the fact that Navier was the first to bring together the scattered knowledge in the field of applied mechanics and strength of materials into a single general theory and showed how many already known laws and methods could be applied in practice to building and the design of structures. He is therefore responsible for creating the discipline of mechanics which we know as structural mechanics." [13] Navier's paper was also available in German and Russian from 1850.

With the theory of trussed frameworks developed almost at the same time in 1851 by Johann Wilhelm Schwedler and Carl Culmann, engineers were now in a position to calculate the precise forces in plane trusses. This new knowledge of structural mechanics was particularly important for the analysis of iron bridges, which were required for the rapid expansion of the railway network. The shortage of iron on the continent of Europe forced engineers to adopt methods of construction that were frugal in their use of materials. "The purpose of every stability analysis, every determination of the forces acting on a structure, is to build the intended construction using the minimum of material," [14] wrote Culmann in 1864 on the role of structural mechanics.

The "time of unsafe tinkering" [15], which had characterised the earlier designs in iron, finally ended in the 1860s. The design of building structures was based on statically determinate systems. "The knowledge that the gulf between the often highly statically indeterminate structures and the still very limited repertoire of methods of analysis is narrowed through the introduction of additional degrees of freedom, the knowledge that more possibilities of movement can mean more calculability, is gradually becoming commonplace." [16]

The graphic statics method developed by Culmann is further developed by contributions from Luigi Cremona (Cremona diagram, 1871) among others into a practical graphical method of structural analysis, which was at its most popular around 1880 and still in use up to 1920. Further work by August Ritter (the method of sections or Ritter's method, 1861), Emil Winkler (the theory of statically indeterminate beams, 1867) and Christian Otto Mohr (Mohr's analogy, 1868) gave a vital boost during this time, which Kurrer defines as the "establishment phase (1850–1875)" in the period of formation of the discipline of structural mechanics. [17] The appearance in 1886 of the book "Neuere Methoden der Statik und Festigkeitslehre" by Heinrich Müller-Breslau containing the process for determining influence lines made it possible for engineers to solve statically indeterminate systems and provided a comprehensive body of theory and tools for the design of structures. Müller-Breslau is therefore referred to as the "consumator of classic structural theory". [18] Although it was at the time possible to solve statically indeterminate systems, they were not used very often in buildings before the beginning of the 20th century because of the large amount of calculations involved. Only for long spans or space frames were statically indeterminate systems used – and then mainly only simple ones – to limit deflections. [19]

The characteristics of structural engineering design and the structural systems used for iron and steel structures between 1870 and 1910 can be summarised as follows:

- Efforts to minimise the amount of calculation with the objective of achieving an efficient design process
- Reliance on plane, statically determinate systems that can be solved easily
- The use to the greatest extent possible of graphical methods for the design of truss systems
- Use of tables to determine the dimensions of individual structural members. In German-speaking countries, for example, these were contained in "Musterbücher für Eisen-Constructionen" by Karl Scharowsky [20] and textbooks by Friedrich Heinzerling.

At this point it should be stated that although the history of structural mechanics, especially the development of the theory of structures, is extensively covered in literature, the history of structural calculations in practice on the other hand has scarcely been explored.

Calculations for Vladimir G. Shukhov's lattice towers

The following sections analyse and evaluate five sets of Shukhov's calculations. The earliest of these calculations stem from the 1911 book by the Russian engineer Dimitri Petrov on water towers in iron, their function, constructions and structural design. [21] The book contains copies of the structural calculations for the water towers in Mykolaiv and Tyumen, which were built in 1907 and 1908 respectively.

Also available are the structural calculations for the Adziogol lighthouse (Figs. 17 and 18, p. 75f.) in Cherson on the Black Sea, which are available in the archive of the Russian Academy of Sciences in Moscow. A condensed version of the original produced by Aleksandr Išlinskij and Irina Petropavlovskaja was translated into German by Ottmar Pertschi in 2003. [22] The considerably more comprehensive full version [23] is hardly legible in parts but contains further important details and drawings. In the same archive

11 Construction work by Bari on the
water tower in Mykolaiv (UA)
1906–1907

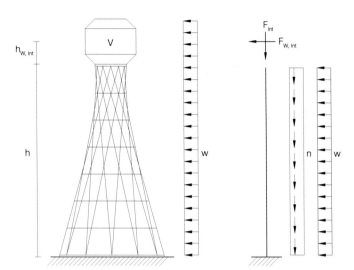

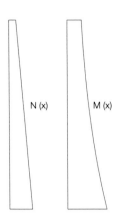

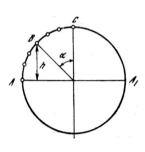

12

13

lies a handwritten fragment of the design of the two-storey water tower in Yaroslavl. The calculations for the NiGRES electricity pylons from the Nizhny Novgorod municipal archives are particularly important. The calculated member forces are also given here to allow a comparison with the analysis of the tower in the chapter "NiGRES tower on the Oka" (p. 96ff.).

Shukhov's method is then extensively explained using the example of the water tower in Mykolaiv. As the calculations all follow the same general approach, only the departures and peculiarities of the four subsequent calculations are highlighted and discussed. To simplify matters, all the Russian units of measurement are converted into the metric system and confusing abbreviations replaced with modern ones in the equations.

Structural calculations for the water tower in Mykolaiv

The water tower designed by Shukhov in 1906 and erected in 1907 in Mykolaiv, Ukraine under the supervision of N. V. Čumakov has an Intze type II tank with a capacity of 50,000 buckets (615 m³). Dimitrij Petrov points out in his book [24] that the water tower in Mykolaiv carried the largest Intze tank in Russia at the time. The hyperbolic lattice mesh is 84 feet (25.60 m) high, the bottom diameter is 42 feet (12.80 m), the top 23 feet (7.01 m). The 48 straight members on the rotation surface are fabricated out of equal-legged angle profiles and vary in size with the height (∑ 5/5/½ inch and ∑ 4½/4½/½ inch, Fig. 15). Almost equal vertical distances (from bottom to top: 2.87 m, 7≈ 2.50 m, 2.41 m, 2.83 m) separate the horizontal rings, which are likewise fabricated out of equal-legged angle profiles. The lattice members are connected by rivets at the intersection points, as are the intermediate rings.

Bari erected the structure in only about six months, from October 1906 to mid-March 1907 (Fig. 11, p. 71). According to Petrov, the members were first bolted together during erection then riveted only after the tower was complete. The tower's outstanding stability was demonstrated during the Second World War when it was blown up by German troops, only to suffer slight damage and be erected again, as was described by J. Belyj. [25]

Design loads and member forces

The calculations use the units of measure common at the time in Russia, e.g. for weight, the pood (1 pood = 16.38 kg), for volume, the bucket (1 bucket = 12.3 litre) and for length, the foot (1 foot = 0.3048 m). A wind pressure of 1 pood/foot² (176.3 kg/m²) is assumed for the design of the structure and applied horizontally or perpendicular to the projected area of the members. Then the horizontally projected area of the four parts of the Intze tank – two cylinders and two frustrums – are determined. In the case of the cone surfaces sloping at the angle α_K – as a rule 45° – the horizontal component of the wind load acting perpendicular to the area is calculated as:

$$F_{Wind, Int} = (\Sigma A_{Cylinder} + \Sigma (A_{Frustrum} \sin\alpha_K)) \, 0{,}667 \, p \qquad (F01)$$

A value of 0.667 was used as the wind pressure shape coefficient for cylindrical surfaces, as his source Shukhov referred to the structural engineer's handbook published by the Akademischer Verein Hütte, Berlin. Using a first moment of area, the distance $h_{W, Int}$ between the top ring of the lattice structure and the resultant force of the horizontal wind load on the Intze tank $F_{W, Int}$ is calculated. The horizontal load p_G per unit of height x acting on the lattice surface is calculated from the number of members 2n and the width b_L of the equal-legged angle sections, with any shielding effect on the members at the sides not being taken into account:

$$p_G = 2 \cdot n \cdot b_L \cdot p \qquad (F02)$$

The loadbearing structure is now divided into ten sections by the nine intermediate rings. Beginning at the top of the lattice structure (x = 0), the moment is calculated based on the height x at each intermediate ring using the expression:

$$M(x) = F_{W, Int} \cdot (h_{W, Int} + x) + \frac{p_G \cdot x^2}{2} \qquad (F03)$$

(Fig. 12). The wind forces acting on the intermediate rings and the central pipe are not considered in the calculation.

Cross section No.	Value μ	l in inches	Cross-sectional area of the angle section: ω in sq. inches	Moment of inertia of cross section	Value φ	Permissible stress in Pood/sq. inch R·φ	Actual stress in Pood/sq. inch Q:ω	Stress reserve in the member in Pood/sq. inch
1	0.00008	9.3 · 12 = 111.6	4.27	7.95	0.65	350.0 · 0.65 = 227	915.44 : 4.27 = 215	12
2	0.00008	94.8	4.27	7.95	0.72	252	222	30
3	0.00008	98.4	4.27	7.95	0.71	248	230	18
4	0.00008	98.4	4.27	7.95	0.71	248	237	11
5	0.00008	98.4	4.27	7.95	0.71	248	244	4
6	0.00008	98.4	4.77	11.10	0.75	262	224	38
7	0.00008	98.4	4.77	11.10	0.75	262	230	32
8	0.00008	98.4	4.77	11.10	0.75	262	234	28
9	0.00008	98.4	4.77	11.10	0.75	262	239	23
10	0.00008	112.48	4.77	11.10	0.70	245	243	2

14

15

Calculating the lattice member forces

To calculate the lattice member forces, Shukhov considers
the tower to be a tube with a cross section that remains plane
and is stiff in shear. The second moment of area about the
axis A-A1 of the cross section (Fig. 13), which is rotationally
symmetrical in plan, is calculated according to Steiner's parallel
axis theorem.

The second moment of area for a total of 2n members is
therefore:

$$I = A \cdot 2 \cdot \sum_{I=1}^{n} (r^2 \cdot \cos^2 i \cdot \alpha) = A \cdot 2 \cdot r^2 \cdot \sum_{I=1}^{n} (\cos^2 i \cdot \alpha) \qquad \text{(F 04)}$$

As the sum of the cosine squared values of α and $(90° - \alpha)$ always
equals one, it follows that:

$$\sum_{I=1}^{n} \cos^2 i \cdot \alpha = \frac{1}{2} \cdot n \qquad \text{(F 05)}$$

which allows F 04 to be simplified to:

$$I = A \cdot r^2 \cdot n \qquad \text{(F 06)}$$

and the section modulus to:

$$Z = \frac{A \cdot r^2 \cdot n}{r} = A \cdot r \cdot n \qquad \text{(F 07)}$$

The maximum lattice member force F_{max} arising from bending is
calculated from:

$$F_{max} = \sigma \cdot A = \frac{M}{A \cdot n \cdot r} \cdot A = \frac{M}{n \cdot r} \qquad \text{(F 08)}$$

In addition to the maximum member force F_{max} due to the bending
moments acting on the tower, the members also carry a force q
due to the weight of the full water tank and a force q_1 representing
the linearly increasing proportion of the self-weight of the tower
structure:

12 Structural model of a water tower with loads and internal forces
13 Determination of the moment of inertia of the tower cross section in
accordance with Shukhov
14 English translation of the calculation tables of the actual and permissible
compressive stress according to Dmitrij Petrov
15 The angle profile size decreases with the height of the water tower for the
All-Russia Exhibition in Nizhny Novgorod (RUS) 1896, which stands today in
Polibino

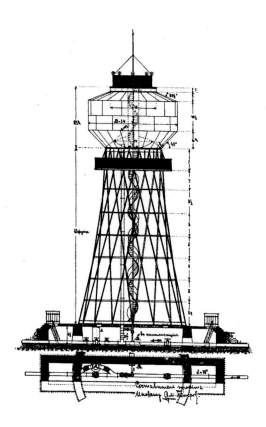

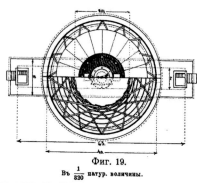

Фиг. 19.

Въ $\frac{1}{320}$ натур. величины.

Желѣзная уравнительная башня съ остовомъ системы инж. В. Г. Шухова
и съ резервуаромъ на 33350 ведеръ воды.
(По проекту инж. Д. В. Петрова).

16

16 Drawing of the water tower in Tyumen (RUS) 1908
17 Adziogol lighthouse, Cherson (UA) 1911

$$q = \frac{V \cdot \rho + m_{WT}}{2 \cdot n} \tag{F 09}$$

$$q_1 = \frac{m_{TG}}{2 \cdot n \cdot h} \tag{F 10}$$

Thus the maximum member force for a given height is:

$$Q = F_{max} + q + q_1\, x \tag{F 11}$$

Calculation of the lattice member cross sections
The buckling stability of the members in compression is then
checked using equation F 12, which Shukhov describes as the
Schwarz-Rankine formula. In reality, this is the formula first published in 1864 by engineers Franz Laissle and Adolf Schübler. They
argued that the widely-used empirical formula devised by Schwarz-
Rankine resulted in unacceptably low safety factors for very slender
members. With the claim to provide the "scientific basis of the
accepted rules" [26], the Laissle-Schübler formula allowed the
permissible extreme fibre stress to be calculated and was therefore
a similar type of formula to Schwarz-Rankine. The Laissle-Schübler
formula is frequently quoted in specialist literature and widely used in
the German-speaking world. In the determination of the maximum
extreme fibre stress, the coefficient φ reduces the permissible stress
in the material. The material itself is taken into account with the
material constant μ, which had been determined in tests (e.g.
wrought iron: $\mu = 0.00008$; cast iron: $\mu = 0.00025$):

$$\varphi = \frac{1}{1 + \mu\,\ell^2\,\dfrac{\omega}{I}} \tag{F 12}$$

The cross sectional area is ω. Shukhov uses the vertical distance
between two intermediate rings as the buckling length ℓ. The actual
lattice member lengths and inclinations, as in the earlier calculation of
the member forces, are not taken into account here. The moment of
inertia I used here is not the one about the weak principal axis but
the moment of inertia calculated about the Cartesian axis. Petrov
says of the calculation of the permissible compressive stresses:
"The formula has a high factor of safety because the value φ for a
column with non-fastened [i.e.: pinned – author's comment] ends is
determined; in the case investigated, one can consider [...] the
individual lengths of an angle profile in the structure as columns with
fastened ends [i.e.: fixed – author's comment]. The actual stress are
therefore considerably less." [27]
The determinant extreme fibre stress is calculated for the various
heights and compared with the permissible extreme fibre stress
according to Laissle-Schübler. The assumed permissible material
stress R is 8.75 kN/cm² (or 350 pood/inch²). Fig. 14 (p. 73) shows the
stresses acting and the permissible stresses for each section.
Similar calculations for the intermediate rings are not present and
the wind loads acting on the intermediate rings are also ignored.

Verification of overall stability
Finally the factor of safety against overturning and therefore the
overall stability of the tower structure is verified. The restoring
moment comprising the effects of the total weight of the iron structure,

the support ring and the empty tank is compared with the calculated maximum overturning moment. The calculations prove only that the restoring moment is greater than the overturning moment – in the case of the Mykolaiv tower the "stability factor" is 1.06. The anchor bolts connecting the support ring to the strip foundation are not taken into account, hence the actual factor of safety against overturning can be assumed to be higher. The structural calculations also contain a verification of the allowable pressure of the strip masonry foundations and a check on the permissible ground pressure.

Structural calculations for the water tower in Tyumen

In 1908, Bari designed a water tower for the municipal water supply in Tyumen, Western Siberia (Fig. 16). The tower has an Intze tank type II with a capacity of 33,350 buckets (410 m³). The hyperbolic lattice mesh is 63 feet (19.20 m) high, the bottom diameter is 40 feet (12.19 m), the top 20.5 feet (6.25 m). The 32 members on the rotation surface are fabricated out of equal-legged angle profiles and change in size once over the height of the tower. Up to the fourth intermediate ring, the angle profiles are 110/110/12 mm and above this level 100/100/12 mm. Seven intermediate rings divide the tower into equal-height vertical sections. The profiles of the intermediate rings decrease in size in 5 mm intervals from the lowest ($\sum 80/80/10$ mm) to the one before the top ($\sum 50/50/10$ mm). The last intermediate ring, which carries a maintenance platform, has the same cross section as the first.

The calculations generally follow the methods used in the design of the water tower at Mykolaiv. In contrast to the earlier example, this tower is analysed and designed using the metric and not Russian system of units. Deviations in the area of applied loads and permissible stresses are also apparent. The design is based on a wind pressure of 180 kg/m² and a permissible steel stress of 1100 kg/cm². The "stability factor" against overturning is calculated as before and equals 1.08. [28]

For the tower in Tyumen, Petrov also calculates a lightness coefficient, which he defines as the ratio of the weight of the structure to the weight of the water supported. In this case, the ratio is 42.80 t/410.55 t = 0.104 and therefore this tower is "more lightweight" than any of the examples discussed by Petrov.

Instead of angle profiles, circular profiles were tried for the tower in Tyumen. However, the required expensive special connections manufactured by the Swiss company gf and their complex installation make the otherwise sensible solution uneconomic, with the result that the idea has to be discarded. [29]

Structural calculations for the Adziogol lighthouse

Fortunately the original "Calculations for a lighthouse with a clear height of up to 68 m according to the system of engineer Šuchov" [30], which relate to the Adziogol lighthouse in Cherson on the Black Sea, have been preserved. The shortened and translated version lacks important deductions in the steps of calculation or presents them incorrectly, therefore the original is also used here to analyse the calculation methods.

The loadbearing structure of the lighthouse (Fig. 18), which is the highest single-storey hyperboloid built by Shukhov, is described as follows: "The lighthouse is a hyperboloid of rotation. The diameter of the bottom at the foundation is 18 m and 7 m at the top ring.

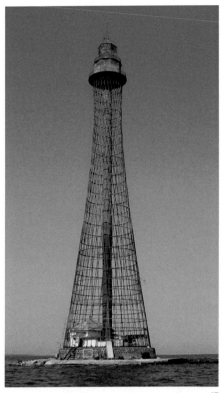

17

18   Adziogol lighthouse, view up the inside of the tower, Cherson (UA) 1911

Design and analysis of Shukhov's tower

The height of the lattice tower from the foundation to the top ring is 59 m. A 4-metre-high metal shack for the lighthouse attendant stands on the surface of the top ring. [...] The lattice consists of 60 angle profiles, which intersect one another and are aligned along straight lines to form a hyperboloid. Several horizontal rings at 2 m vertical spacings are fastened from the inside of the tower to the angle profiles of the hyperboloid to create the necessary connections between them. The angle profiles are riveted together at the intersection points and as a result the lattice tower is a composite and rigid system. An iron pipe with a diameter of 2 m, which contains an iron spiral staircase, runs through the middle over the entire height." [31]

With this structure too the calculations follow the same basic method used for the earlier water towers at Mykolaiv and Tyumen. However the calculations are considerably more comprehensive and more detailed than the others. For instance, the calculation processes that would otherwise be used without comment are deduced and explained.

Derivation of the wind loads
Particular focus is placed on the derivation of the wind loads acting on the structure – undoubtedly due to the exposed location by the sea. The wind pressure is assumed to be 275 kg/m^2. For the equal-legged angle profiles, the normal and then the necessary force component acting in the direction of wind flow are calculated taking into account the position in plan (Fig. 21 a, p. 79). To do this, the angle profiles were assigned to one of four sections according to their position in the circular plan shape (Fig. 21 b, p. 79). Depending on the direction of wind flow, the projected width of the angle is first calculated, then the component normally acting on this width, and finally the force component acting in the direction of wind flow. If a leg runs parallel to the direction of the wind, the resulting force is equal to the wind pressure (p) times the leg length (a). In the region A to B, the resulting force is much less, at 45° only 0.353 pa (point B). Between B and C, the back leg is shielded by one at the front; as angle α increases, the resulting force increases again to pa. The resultant force is greater than pa only between points C and D, because here the angle profiles "catch" the full pressure of the wind, the maximum is reached at 135° if the open v of the angle profile points downwards. The maximum force is 1.414 pa here. The paper also goes on to comment that the region from D to E should be treated like B to C and the region E to F like A to B. The resultant in the region between F and A is calculated as: pa (sin^3 α + cos^3 α), contrary to the original writing error in the drawing (in Fig. 21 b, p. 79 corrected). The minimum resultant in this region with α = 315° is 0.707 pa.

As an example, if the applied wind pressures are estimated for a tower with 48 members, the direction of wind flow varies in 15° divisions. The total wind force is 42.258 pa, from which an average of 0.88 pa is applied to each member.

If this approach is applied to the structure of the lighthouse and the wind direction varied for the 60 members in steps of 6°, the total force is 53.562 pa or 0.893 pa per member (not included in calculations). The example helps to explain the assumption of an equal horizontal wind load pa for all vertical members, irrespective of the direction of the wind. Because the resultants on the angle profiles

are less than pa in three quarters of the circle and greater only between C and D, it is argued that the assumption of a uniform horizontal wind load pa – which influences the stability of the entire structure – lies on the safe side.

According to the structural calculations, the angle profiles of the vertical members are stepped down in size slightly every 10 m (except the first butt joint after 9 m), from 88/88/12 mm to 88/88/10 mm, 76/76/10 mm, 63/63/10 mm to 50/50/10 mm. The wind force acting on the lattice mesh was therefore calculated in this way. The wind force on the pipe was then calculated and added to the total. This is commented upon in Shukhov's calculations: "The wind pressure on the pipe as a cylinder (normally the wind pressure on a cylindrical surface is assumed to be p_o = 0.45 pdh to 0.57 pdh, compared to Hütte, p. 276, part 1, 6th edition) where p_o = 2/3 p = 180 kg per linear metre in vertical projection." [32]

$$p_G = p\left(n \cdot b_L + \frac{2}{3} \cdot d\right) \qquad \text{(F 13)}$$

The calculations go on to say: "The size of the actual wind pressure on the lattice is of course smaller than the calculated value because the angle profiles that lie in the segments in the tangential regions relative to the wind direction overlap one another and therefore considerably reduce the surface acted upon by the wind." [33] The moment at a height i of the tower is calculated using a slightly modified version of expression F 03 (p. 72) including the additional top moment as:

$$M_i = M + Q(x) \cdot x + \frac{p_G \cdot x^2}{2} \qquad \text{(F 14)}$$

The critical member forces and the resulting stresses in the members are calculated below at various heights and expressed in tabular form (Figs. 19 and 20, p. 78). The maximum resulting member forces arising from the applied bending moment are calculated from the equation given earlier using the section modulus of the tower (see F 06, p. 73). The compression and tension forces in the members are calculated by superimposing the effects of the applied loads and the self-weight of the tower at the various heights. In contrast to all the other calculations, this design takes into account the bending of the lattice members under a wind load of pal^2/24. Why Shukhov uses this value, which is considerably less than the support moment of a multispan continuous beam, is not apparent. As in the earlier examples, the check of the members in compression uses the Laissle-Schübler formula. To calculate the critical compressive stress, the compressive force is divided by the area of the gross cross sectional area and superimposed with the effects of bending, while the permissible tensile stress check is based on the net cross sectional area (Fig. 20). The permissible stress is assumed to be 1000 kg/cm^2.

Comparative calculation using a tubular cross section
Shukhov performed a comparative calculation of the overturning moment for a hyperboloid of rotation with a solid surface. The

19

Таблица 1

Q, m	M, t.m	R, m	R, n	p, m	S, m	T, m	w, cm² brutto	netto
10	4,0	3,5	105	0,381	0,881	−0,119	—	—
16	105	3,35	100	1,04	1,01	+0,47	8,857	7,257
22	200	3,3	99	2,02	2,65	+1,39	—	—
29	327,5	3,5	105	3,12	3,84	+2,40	11,277	9,677
36	490	3,8	112	4,30	5,10	+3,5	—	—
44,1	690,25	4,3	133	5,35	6,25	+4,45	13,696	12,096
52,2	931	4,9	147	6,33	7,33	+5,33	—	—
60,3	1212,25	5,5	165	7,34	8,45	+6,23	17,87	15,59
69,4	1534	6,25	187	8,13	9,41	+6,95	—	—
77,5	1898,75	7	210	9,04	10,41	+7,67	18,657	16,377
86,6	2309	7,7	231	10	11,5	+8,5	—	—
94,8	2717,15	8,4	252	10,78	12,38	9,18	21,141	18,881
103	3158,6	9	272	11,65	13,35	10,61	—	—

φ	M_b, kg·cm	W, cm³	T/W_netto	T/W_brutto	M_b/W kg/cm²	Σ напряжение сжатия kg/cm²	Дополнительное напряжение
0,41	230	5,479	65	182	42	224	410
—	—	—	191	300	42	342	416
0,53	290	9	248	341	33	374	531
—	—	9	362	452	33	485	540
0,63	350	13,36	357	457	27	434	630
—	—	—	440	535	27	562	635
0,6	350	17,03	400	473	20	493	640
—	—	—	445	527	20	547	600
0,7	405	21,22	468	558	19	577	780
—	—	—	520	616	19	635	707
0,7	—	23,78	488	586	18	604	701
—	—	—	538	635	18	653	710

20

projected surface of the hyperboloid is simplified into separate cylindrical and frustrum sections. The same wind pressure of 180 kg/m² assumed for the cylindrical pipe section containing the staircase is used here. The total overturning moment is 3323 tm, a value which is only slightly greater (by about 5 %) than the value calculated for the lattice structure.

Influence of the intermediate rings

A further section considers the intermediate rings; it is the only comment about them found in the calculations analysed here. Shukhov writes: "The purpose of the horizontal intermediate rings is to hold the intersection points of the angle profiles of the tower in position and distribute the wind pressure acting on the angle profiles evenly through the intersection points" and continues: "The greatest thrust in the ring ab to be accommodated by the lattice can, like the half wind pressure on the angle profiles of the lattice with a height h or the thrust on the ring, be assumed to be:

$$Q = \frac{p \cdot a \cdot 2n \cdot h}{2}$$ (F 15)

[here h is the distance between two intermediate rings – author's comment]." [34]
The compressive stresses on the different cross sections of the intermediate rings, which vary in a similar way to the vertical members in different parts of the tower, are calculated from the force Q. The basic model is a horizontal ring supported at opposite points

(Fig. 22). The maximum force Q is taken out of the rings at these points and transferred into the lattice structure. The resulting horizontal forces of both sides of the ring, compression and tension, cancel one another out at the support points.
Surprisingly, no check of the permissible compressive buckling stress using the Schwarz-Rankine or Laissle-Schübler method is carried out, although the achieved stress is – from consideration of the lowest intermediate ring – greater than the permissible steel stress of σ = 1000 kg/cm² found elsewhere in the calculations (Fig. 25). Shukhov goes on to comment: "The actual pressure is considerably smaller because the connection points 00₁, which transfer this pressure, also transfer it at connection points cc etc. into the lattice. [...] The rings at any point over the full height of the tower are made from the same size of angle profile as that of the lattice at that point." [35] (Fig. 24)

Spoke wheels

In the vertical central axis of the tower is a fabricated pipe for the spiral staircase. The diameter of the pipe is 2 m, the plate is 5 mm thick. The pipe cross section is restrained at 10 m centres by radial tension rods connected to the loadbearing structure (Fig. 18, p. 76). The pipe is designed as a continuous beam; the maximum support moment stress in combination with the compressive stress from self-weight is compared to the permissible material stress. The two lower spoke wheels consist of 32 tension rods, the 3 upper spoke wheels have 16. However, the design of the spoke wheels will not be discussed further here.

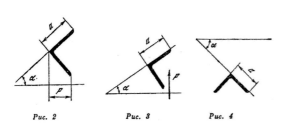

Puc. 2   Puc. 3   Puc. 4

a

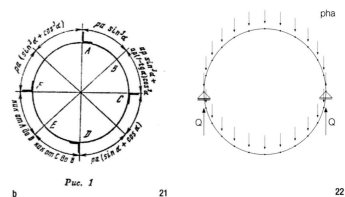

Puc. 1

b   21   22

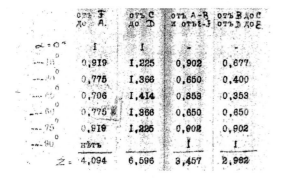

23

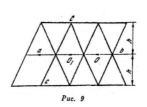

Puc. 9

24

Таблица 3

Уголок	Q, m	w, см³	σ, кг/см²
2 × 2 × 3/8″	16,5 × 0,05 = 0,825	7,3	112
2 × 2 × 3/8″	16,5 × 0,063 = 1,04	9,7	107
3 × 3 × 3/8″	16,5 × 0,076 = 1,25	12,1	104
3 × 3 × 3/8″	16,5 × 0,088 = 1,45	16,3	90

25

The structural calculations for the two-storey water tower in Yaroslavl

The water tower in Yaroslavl consists of two segments with heights of 19.20 m (bottom) and 20.27 m (top) (Fig. 26 b, p. 80). The two suspended bottom tanks have a capacity of 116.6 m³ (bottom) and 194.2 m³ (top). The cross sections each consist of 30 vertical members, which are 4.5/4.5/5/8 inches to 4.5/4.5/1/2 inches in the bottom section and 4.0/4.0/1/2 inches to 3.5/3.5/3/8 inches in the top. Seven horizontal intermediate rings stiffen each segment. The calculations for this structure, which are limited to two and a half handwritten pages (Fig. 26 a, p. 80), also follow the approach set out above. In the design of the lower segment, the bearing forces from the upper section become the applied loads, without further details of how they are transferred. The main ring, which separates the two sections, is not included in the calculations. The basic assumptions for loads and material properties are: material stress depending on the component: 350 and 400 pood/inch² (890 kg/cm² and 1015 kg/cm²), wind pressure: 0.7 pood/foot² (123.4 kg/m²). However, the principal dimensions upon which the design is based differ greatly from those of the as-built tower, which suggests that the calculations refer to an earlier preliminary feasibility study.

Structural calculations for the NiGRES tower on the Oka

Only very few of Shukhov's original calculations for multistorey towers are still in existence. Fortunately, the historical calculations for the NiGRES tower emerged from the city archives at Nizhny

19 Bending moments and transverse forces acting on the tower
20 Calculation tables for member forces and permissible/actual stress checks: in the bottom table, the penultimate column shows the combined compressive stress calculated by adding the stresses due to bending and direct compression; the last column shows the permissible compressive stress.
21 Wind flow around the angle profiles for variously oriented members (a) and according to their position in plan (b), error in original formula corrected
22 Reconstruction of Shukhov's assumed model for assessing the action of the intermediate rings
23 Sum of the projected member surface areas depending on wind direction
24 Schematic drawing of the action of the intermediate rings
25 Maximum effective force in the intermediate rings according to equation F 15 (p. 78)

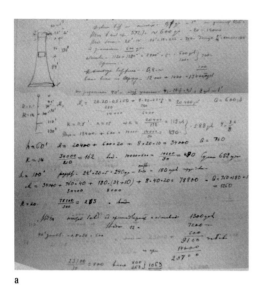

a

26 Water tower in Yaroslavl (RUS) 1911
a First page of Shukhov's design
calculations
b Photograph ca. 1911
c Blueprint (different to the built
version)

b

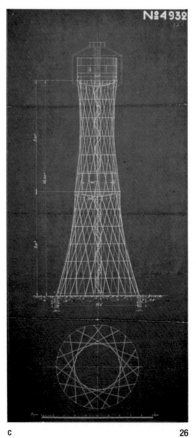

c   **26**

Novgorod (Fig. 27, p. 82). [36] The calculations contain the load assumptions for the straight members of each section, including the design of the riveted connections and the ring foundation. All this is expressed in only ten typewritten pages (Fig. 27, p. 82f.). Whether this document contains all the calculations Shukhov produced for the tower cannot be confirmed. It is possible that they are a short summary, which he may have had to submit to the local authority before building could start. In any case, the document provides a rare and revealing insight into the structural engineering concept and design of the multistorey tower and the underlying structural model and load assumptions. [37]

Load assumptions and member forces

At the beginning of the calculation, it says without any further explanation that according to "technical agreements", of which no details are given, the maximum loads occur when "the conductors are intact, no ice load is present and the wind pressure is 250 kg/m²". [38] Clearly Shukhov relates here to different load case combinations which had to be investigated according to the standards of the time. The above-mentioned wind pressure of 250 kg/m² is constant and applied over the full height of the tower. For the straight members, the area subject to wind pressure used in the calculation is the leg length of the equal-legged steel angle profiles.

The design of this tower differs from the previous calculations in that it also takes into account the wind load on the horizontal intermediate rings. The shielding effect at the flanks of the structure are taken into account in the calculation of the wind loads on the intermediate rings. The resulting wind force on an intermediate ring with a diameter D and a leg length of the angle profile of b is calculated using the expression:

$$W_{Ring} = 250 \text{ kg/m}^2 \cdot \frac{4}{3} \cdot D \cdot b \qquad \text{(F 16)}$$

Compared with the projected length of D per semicircle, this represents a reduction of 33 %. Interestingly, Shukhov uses the same reduction factor here that was found for the Adziogol lighthouse constructed in 1911 in Cherson. In the associated structural calculations, he refers to the engineers' handbook of the Akademischer Verein Hütte, which was an engineers' reference book published in Berlin in 1857. In the 1906 edition, the wind load acting on a cylinder with a diameter d and height h in the direction of the wind is calculated with the formula $W_{Ring} = \frac{2}{3} \cdot d \cdot h \cdot p$. [39] For the open-sided annulus exposed to wind, Shukhov therefore uses this formula times two to simulate both halves of the annulus. The quoted source of information is one of many examples showing that Shukhov studied international specialist literature and put this knowledge of contemporary research to good use in his projects.

The self-weight of the tower including the conductors is given as 144.26 t. The overturning moment for the whole tower is calculated from the resultants of the wind forces calculated for each section. The resultant wind force on each section is always applied at half height. From the original calculations, the overturning moment is

7384.21 tm (tonne metre). The maximum compressive forces arising from the bending moment were calculated for a number of member pairs n and radius r as before using formula F 08 (p. 73). From the above self-weight and overturning moment, a maximum compressive force for the first section is calculated as:

$$F_{D, max} = 144.26 \text{ t}/40 + 7384.21 \text{ tm}/(20 \cdot 17 \text{ m}) =$$
$$3.6 \text{ t} + 21.7 \text{ t} = 25.3 \text{ t}$$

From the same values, the maximum tensile force is calculated as:

$$F_{Z, max} = 7384.21 \text{ tm}/(20 \cdot 17 \text{ m}) - 144.26 \text{ t}/40 =$$
$$21.7 \text{ t} - 3.6 \text{ t} = 18.1 \text{ t}$$

Design of the lattice members

Using this compressive force, the member stresses in the vertical members in the first section are calculated. The buckling length is taken as the vertical distance between two horizontal rings. For the design of the vertical members, Shukhov did not use the Laissle-Schübler formula as in the earlier towers, but rather the calculation method published in 1906 by Johann Emanuel Brik [40] with a safety factor of 2.5. The permissible compressive stress according to this method is:

$$f_{c, max} = 0.4 \cdot (3100 - 11.4 \cdot \lambda) \qquad \text{(F 17)}$$

with a slenderness factor $\lambda = l_c / i_y$. Interestingly, he does not use the smaller radius of gyration i_ζ about the principal axis, but rather the radius of gyration i_y about the Cartesian axes. The explanatory notes do not make it clear whether buckling about the strong axis is considered, because "the feet of the tower are considered as a column restrained at both ends" [41]. Clearly Shukhov assumes that the support of the angle profiles can be taken as torsionally stiff and buckling about the weak axis can therefore be excluded. A hypothesis which, at least as it affects the connections at the intermediate rings, is worthy of discussion today.

For the maximum compressive force calculated above for the first section, the buckling check on the members is carried out according to Brik. The permissible stresses are:

vertical members: $\Sigma 12/12/1.2$ cm mit $i_y = 3.65$ cm,
$A = 27.54$ cm², $l = l_c = 244$ cm $\int \lambda = 67$
$\int f_{c, max} = 0.4 \cdot (3100 - 11.4 \cdot 67) = 934$ kg/cm²
$\int f_c = 25{,}300$ kg/27.54 cm² $= 920$ kg/cm² $< f_{c, max}$

The calculation of the overturning moment is repeated for each section to design the members for the maximum compressive force. Finally the riveted connections for the critical members were designed in each section. The calculated rivet diameters are ⁷⁄₈, ³⁄₄ and ⁵⁄₈ inch.

Design of the foundation

Anchor bolts transfer the calculated maximum tensile forces into the foundation. The strip foundation of lightly reinforced concrete is designed for the calculated lifting tensile force of $F_{Z, max} = 18.1$ t.

Page 1

РАСЧЕТ БАШНИ ВЫСОТОЙ 128 МТР.

сист. инж. В.Г.Шухова для линии электропередачи
Растяпино-Богородск 115 кв. Промежуточная мачта № 3
черт. № 9568.-

Наибольшие усилия в элементах башни получаются
для 1-го случая нагрузок,указанных в технических условиях
для расчета опор,а именно для случая необорванных проводов,от-
сутствие гололеда и давления ветра 250 кг/кв.мтр.

Для рассматриваемого случая из технических
условий имеем:

давление ветра на провода 1725х3 = 5175 кг. собственный вес
проводов и изоляторов (1120+2 х 50) 3 = 3660 кг.

Давление ветра на башню по черт. № 9568 имеем:
на верхушку (см. черт. 9572)
250(6,2х3х0,075+12х4х0,038+5х0,2) = 2100 кг.

На V секцию.
на стойки на L 102 x 102 x 9,5
250х 20 х 24,8 х 0,1 = 12400 кгр.
на кольца по ф-ле:
250х 4/3 · Dх, где
D - суммарный диаметр колец
ℓ - ширина кольца

Верхнее кольцо секции из L 75 х 75 х 8
D = 6 мтр.   ℓ = 0,075
Промежуточные кольца   из 50 х 50 х 6
D сумм = 62,6   ℓ = 0,05
250 х 4/3 (6х0,075 + 62,6 х 0,05) = 1190 кгр.
Общее давление ветра на V секцию
R_V = 12400 + 1190 = 13590 кгр.

На IV секцию
На стойки на L 102 x 102 x 12,7
250х20х25,5 х0,1 = 12750 кгр.
Верхнее кольцо секции   на L 75 х 75 х 8
D = 10 мтр.   ℓ = 0,075

Page 2

- 2 -

Промежуточные кольца   на 63 х 63 х 6,35
D_ср = 110,8 мтр.   ℓ = 0,053
250 х 4/3 (10х0,075 + 110,8 х0,053) = 2580 кгр.
R_IV = 12750 + 2580 = 15330 кгр.

На III секцию.
На стойки их L 102 x 102 x 9,5
250х 40 х 25,7 х 0,1 = 257000 кгр.
Верхнее кольцо секции из L 75 х 75 х 8
D = 14 мтр.   ℓ = 0,075
Промежуточные кольца из 75 х 75 х 8
D сумм = 159,7 мтр.   ℓ = 0,075
250 х 4/3 . 173,7 х 0,075 = 4340 кгр.
R_III = 25700 + 4340 = 30040 кгр.

На II секцию.
На стойки их L 102 x 102 x 12,7
250х40х26х0,1 = 26000кгр.
Верхнее кольцо секции из L 90 х 90 х 10
D = 19,4 мтр.   ℓ = 0,09 мтр.
Промежуточные кольца из L 75 х 75 х 8
D_ср = 217,98 мтр.   ℓ = 0,075
250 х 4/3 (19,4 х0,09 +217,98х0,075) = 6030 кгр.
R_II = 26000 + 6030 = 32030 кгр.

На I секцию.
На стойки на L 120 x 120 x 12 мм.
250 х40х26,8х0,1 = 32160 кгр.
Верхнее кольцо секции на L 102 x 102 x 12,7
D = 25,8 мтр.   ℓ = 0,1
Нижнее кольцо секции на L 120 x 120 x 12
D = 34 мтр.   ℓ = 0,12
Промежуточные кольца на L 80 x 80 x 10
D_ср = 288,56 мтр.   ℓ = 0,08

Page 3

- 3 -

250 х 4/3 (25,8 х 0,1 + 34х0,12+288,56х0,08) = 9910 кгр.
S_I = 32160 + 9910 = 42070 кгр.

Вес башни
Верхушка   -   4400 кгр.
V секция   -   10300 -"-
IV -"-   -   14900 -"-
III -"-   -   24400 -"-
II -"-   -   33400 -"-
I   -"-   -   53000 -"-
Опорные части   -   5100 -"-
                   145700 кгр.

I Секция.
Ломкий момент башни относительно низа
M = 5,175х130х2,1 +127 +13,59х111,75+15,33х97,15 + 30,04х
х52,35+32,03х37,35+42,07х12,45 = 7384,21 т/мтр.
Усилие на стойку от давления ветра их по ф-ле
U = M/(r·n)
где   r - радиус окружности рассматриваемого сечения
       n - число стоек сечения
Для рассматриваемого сечения
r = 17 мтр.   n = 20
U = 7384,21/(20 х 17) = 21,7 тонн.
Усилие на стойку от веса конструкции
C_u = 1/40 (145700 - 5100 +3660) = 3600 кгр.
Суммарная нагрузка на 1 стойку
C_I = C_U + C_u = 21,7 + 3,6 = 25,3 тонн.
Сечение стойки L 120 x 120 x 12 мм.
Площадь сечения уголка w = 27,54 ст.кв.
Радиус инерции сечения   r = 3,65 ст.
Радиус инерции уголка берется относительно оси
параллельной полке и проходящей через центр тяжести уголка,так
как уголки башни,составляющие ноги её в отношении сопротивления
сжатия рассматриваются как стойки,закрепленные двумя концами.

Page 4

- 6 -

C_u = 57,35/40 = 1,5 тонн.
Суммарное усилие
C_u = 12,2 + 1,5 = 13,7 тонн.
Сечение стоек L 102 x 102 x 9,5
Площадь сечения уголника w = 18,5 ст.
Свободная длина ℓ = 233 ст.
Радиус инерции r = 3,1 ст.
Отношение ℓ:r = 233 : 3,1 = 75
и допускаемое напряжение
К доп. = 0,4 (3100 - 11,4 х 75) = 898 кгр/ст.
действительное напряжение
К действ. = 13700 : 18,5 = 740 кгр/ст.
Заклепки Д-4" , число срезов n = 8
К = 610

IV. Секция.
Ломкий момент относительно низа секции
M = 5,175 х55,3 + 2,1 х 52,3 + 13,59 х37,05+15,33 х 12,45 = 1090,4 т/мтр.
радиус окружности r = 7 мтр.
n = 10
C_u = 1090,4/(7 х 10) = 15,6 тонн.
от веса конструкции
C_u = 93,26/20 = 1,7 тонн.
C сумм = 17,3 тонн.
Сечение стоек L 102 x 102 x 12,7
Площадь сечения стойки w = 24,4 ст.
ℓ = 232 ст.   r = 3 ст.   ℓ:r = 78
Допускаемое напряжение
К доп. = 0,4 (3100 - 11,4 х78) = 884 кгр/ст.
Действительное напряжение
К действ. = 17300 : 24,4 = 710 кгр/ст.
Заклепки   Д-4"   n = 8
К_r = 735 кгр/ст.

Page 5

- 7 -

I. Секция.
Ломкий момент относительно низа секции
M = 5,175х 30,6 + 2,1 х 27,6 + 12,59 х 12,35 = 380 т/мтр.
Радиус окружности r = 5 мтр.
n = 10
U = 380/(5х10) = 7,6 тонн.
От веса конструкции
C_u = 18,36/20 = 1 т.
C сумм. = 8,6 тонн.
Сечение стоек L 102 x 102 x 9,5
Площадь сечения w = 18,5 ст.
ℓ = 348 ст.   r = 3,1 ст.
Допускаемое напряжение К доп. = 04(3100 - 11,6х90)=875кгр.
К действ. = 8600 : 18,5 = 465 кгр/ст.
Заклепки Д- 4" n = 2 К = 610 кгр/ст.

Верхушка башни. Черт. № 9572

Верхняя балка:
Наибольшее усилие на балку имеет место для случая
II,когда провода покрыты слоем гололеда.
Вес проводов с гирляндами в этом случае равен
1940 х 3 = 5820
Ломкий момент балки
M = 1940 х 160 = 310.000 кгр/ст.
Сечение балки № 20
Момент сопротивления W = 191 х 2 = 382 ст.
и напряжение материала балки
К = 310000 : 382 = 810 кгр/ст.
Наклонные тяги.
Наибольшие усилия в тягах получаются для случая I.
Давление ветра на провода 5175 кгр.
Давление ветра на верхушку 2100 кгр.

Page 6

- 8 -

Усилие растягивающее тягу
(5175 + 2100/2)·12,54/5,2 + √(12,54² + 2,6²)/12,54 = 7650
Сечение тяги L 75х75 х 8 мм.
w_1 = 11,47 ст.   w_пол = 9,87 ст.
К = 7650 : 9,87 = 775 кгр/ст.
Заклепки Д -4" ,число срезов n = 4
К = 630 кгр/ст.
Наклонные подкосы к верхней балке.
Наибольшие усилия в наклонных подкосах имеют место
для случая I
Усилие сжатия подкоса
6225 · 8,4/2,6 · 1/2 + √(8,4² + 2,6²)/8,4 = 5270 кгр.
От вертикальной нагрузки проводов
1/2 1220 · 7,5х0,4 х8,8/5,9х6,1х0,4 = 1130 кгр.
Суммарное усилие сжатия
5270 + 1130 = 6400 кгр.
Свободная длина стержня ℓ = 440 ст.
Сечение L 75х75 х 8
Площадь w_1 = 22,94 ст.   w_пол = 20,38 ст.
Радиус инерции r = 2,85 ст.
440 : 2,85 = 155
Допускаемое напряжение
К = 0,4 БП (-r/ℓ)² = 353 кгр/ст.
Действительное напряжение
К действ. = 6400 : 22,94 = 280 кгр/ст.
Заклепки Д- 5/8" ,число срезов n = 6
К = 535 кгр/ст.
Подкос вертикальной плоскости
Усилие в подкосе.

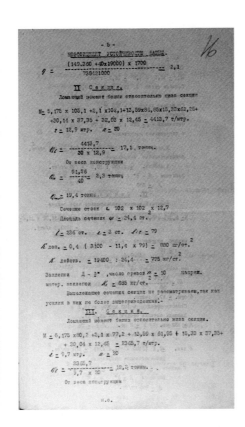
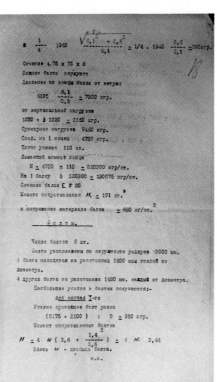
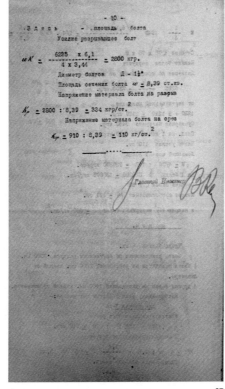

27 Shukhov's structural calculations for the NiGRES tower on the Oka, Dzerzhinsk (RUS) 1929, undated original document

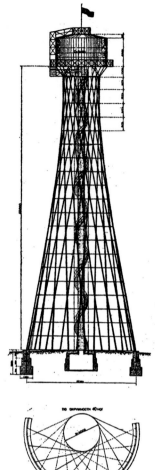

Рис. 1. Водонапорная башня
системы инженера В. Г. Шухова высотою 45 м, резервуар
емк. 20 000 ведер

a

b   28

In the calculations, the ring foundation is 280 cm deep and 180 cm wide and has $4 \approx 70$ cm steps . Assuming a density of 2.0 t/m^3, the proportion of the foundation assigned to one vertical member is calculated as having a weight of 19 t. Unlike the one in the calculations, the as-built strip foundation has a trapezoidal cross section, with dimensions that deviate from those in the structural calculations.

The calculations also contain the design of the anchor bolts and the steel beams forming the extensions at the mast head. The calculations do not contain a design for the horizontal truss members which divide the tower into its individual sections.

Evaluation of the historical calculations

The structural engineering calculations for the towers are analysed in the following section on the one hand from the view of Shukhov's Russian contemporaries, in particularly those who had been involved with the collapse of a water tower in Dnipropetrovsk, the structural behaviour of which has already been analysed (see "Water tower in Dnipropetrovsk (1930)", p. 60ff.), and on the other hand from today's point of view.

The collapse of the water tower in Dnipropetrovsk

A water tower of a Shukhov-type design was erected at the Karl-Liebknecht factory in Dnipropetrovsk, Ukraine in August 1930 (Fig. 28a). With a tank volume of 250 m^3 and a tower height of 45 m, this structure was the tallest single-storey Shukhov water tower of its time. Only a few days after its erection, the tower collapsed under what could be described, at worst, as moderate wind conditions (Fig. 28b). As the site supervision records confirm that the structure was erected in accordance with the requirements and that the steel was of the required quality, this incident throws up questions about the stability of the structure. The collapse was analysed by a specially convened investigatory commission at which the stability and constructional details were discussed by numerous experts.

The most important document on the collapse of the tower is the paper "Collapse of a water tower" [42] published in 1931 by the local consultant engineer V. A. Djadjuša, who was a member of the investigatory commission.

According to this source, the structure failed in the middle of the tower shaft first where it deformed "like a piano accordion" and then collapsed. The rest of the structure remains within the ring foundation, which points towards a stability failure. Shukhov's design of the tower, which is gone through briefly in the paper, generally follows the already discussed method used for the other water towers. In his concise analysis, Djadjuša cites three main reasons for the collapse of the tower:

- Serious shortcomings in the design: he criticised in particular what he believed to be the weakness of the node points (connection of the verticals to the intermediate ring) and the connection of the verticals to the bottom support ring.
- The structural calculations appeared insufficient and in his view needed to be completely revised. He goes on to criticise the buckling check: instead of the vertical members having fixed supports, they should be assumed to be pinned. He refers to the use of Cartesian axis of the angle profiles instead of the weak

principal axis for the determination of the radius of gyration.
• Based on the Soviet and German construction standards of the
time, he points out in particular that the assumed wind loads
appear to be much too small and should be increased by at least
25 %. The factor of safety against overturning is also considerably
too low.

Djadjuša comes to the conclusion that Shukhov's hyperbolic lattice
towers should be used only for low or medium-height water towers,
because they are not stiff enough for greater heights or heavier
loads.

Historic investigations into Shukhov's lattice towers in Russia
Prompted by the collapse of the water tower in Dnipropetrovsk and
the widespread use of Shukhov's hyperbolic lattice towers, Russian
scientists and engineers concerned themselves on many occasions
over the following decades with the structural behaviour and the
design of these structures.

Analysis of the structural behaviour by G. D. Popov (1931)
Popov [43] calls into question the assumption of a shear-stiff tubular
cross section for the design, as this is valid only for shells or appropriately meshed prefabricated tubes but not for the lattice surface
formed here using straight members. In contrast to the situation with
truss or triangular structures, the transfer of shear forces at the intersection points of two members carrying normal forces cannot be
guaranteed (Fig. 30).
Assuming flexurally stiff intermediate rings, which, unlike the
Shukhov structures, coincide with the intersection points of the verticals, he proposed an alternative approximate method of design.

Stability calculations by Aleksandr Nikolaevich Dinnik (1950)
Also with reference to the collapse of the tower in Dnipropetrovsk,
Russian engineer Dinnik derived a formula for the approximate
determination of the critical load of a hyperbolic lattice structure in
his paper "Stability of curved lattices" [44] published in 1930. His
key conclusion is that in determining the buckling load, not only

29

28 Water tower, Dnipropetrovsk (UA) 1930
    a Elevation and plan
    b Photograph of the collapsed tower
29 NiGRES tower on the Oka, Dzerzhinsk (RUS) 1929
30 Transfer of transverse forces

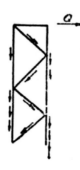

30

should the buckling failure tangential to the intermediate rings be taken into account but also the torsional effects, i.e. torsional-flexural buckling. Using the formula he developed for framework structures and a few simplifications, he calculated a critical load of 204 t for the water tower in Dnipropetrovsk (tank capacity 250 m³). A value well below the calculated load capacity quoted in the section "Water tower in Dnipropetrovsk (1930)" (p. 60f.).

Model studies of B. V. Gorenšteijn (1959)

Gorenšteijn carried out a large number of model studies of hyperbolic lattice towers. [45] Fig. 31 shows a model with diameter of 300 mm and 36 member pairs (length 250 mm, diameter 3 mm) and three intermediate rings (diameter 2 mm), which are arranged at the welded intersection points.

The photographs show that the lattice members bend in two half-waves and intermediate rings in multiple waves. Gorenšteijn also stated that the model structure failed not through reaching a general stability limit but by the connections coming apart and therefore the results of the model investigations were not conclusive. That said, the tests clearly show the global buckling of the lattice.

With the results of the tests, Gorenšteijn proves wrong one of the calculation methods proposed in the 1930s by N. P. Griškova [46] and others, which is not elaborated upon further here because of its irrelevance and Gorenšteijn's proof of its shortcomings. In spite of the above-mentioned inadequacies of the models, the load capacities are still around four times the load capacities calculated according to Griškova.

Other analyses by Russian scientists

In the 1950s further analyses of Shukhov's towers were performed in Russia. The analyses sought to assess the critical loads of the structures by means of complex mechanical formulas. They included the investigations by Leonid Samuilovič Lejbenzon (1951), N. A. Kolkunov (1959) and A. M. Maslenikov (1959). The formulas were often more than half a page long and therefore unwieldy and impractical. Seen from today's point of view, they appear no longer relevant for the structural design of towers; their analyses are therefore not discussed further in this book.

A modern evaluation of Shukhov's structural models

The earlier sections of this chapter show the analyses and evaluations of the structural models and verifications of examples of Shukhov's towers designed between 1906 and 1930. This material provides a randomised source for discovering the methods and development of his calculations.

In general, it can be concluded that the main method used to design and model these structures did not change over almost 25 years between the water tower in Mykolaiv and the NiGRES tower. And it is very probable that the same model and method of design were used in the ten years between 1896 and 1906 as well. Hyperbolic lattice towers are always considered as shear-stiff tubular cross sections of variable diameter. The section modulus is calculated at various heights according to the assumptions of the Bernoulli hypothesis (plane sections remain plane, no shear deformations) and the bending moment used to calculate the resulting maximum member force. The design of the members is then based on the determinant member force at the various heights. This is the reason why the member cross sections even for single-storey towers are in a graduated range of sizes, sometimes two or more, over the height of the tower. The calculations contain a series of simplifications which seem astonishing from the viewpoint of today. The most significant are:

- As demonstrated in the section "Horizontal load transfer" (p. 34ff.), the assumption of a shear-stiff circular cross section for the calculation of the maximum member forces is inadequate because the forces in the vertical members are often seriously underestimated for rotation angles that deviate from 90°.
- The inclination of the member is generally not taken into account in the calculations. It does not appear in the calculation of the true member lengths to arrive at buckling length of the members, nor are the vertical forces (from the top load or the bending moments resulting from wind loads) appropriately calculated. Hence the member forces and buckling lengths upon which the calculations are based are too small.
- The five towers discussed above have a loadbearing lattice of equal-legged angle profiles. The moment of inertia about the Cartesian axis is always used instead of the weaker value about the principal axis of the angle profile in the check on buckling. Shukhov argues that buckling of the angle profiles could be expected to take place only in this direction because of the connections between them and the horizontal ring. However, this assumption is false because of the pinned design of the node points between the verticals and the intermediate ring, as can be seen in the various failure shapes in the section "Mesh variant 2: Construction used by Vladimir G. Shukhov" (p. 52ff.).
- Wind loads acting on intermediate rings are not taken into account in the calculations except for the NiGRES tower. In the case of the latter, a reduction factor to adjust for the wind resistance of solid cylinders given in engineer's reference books is applied to the loads. The transfer of this factor to open rings of small cross sectional size seems questionable.
- In the consideration of the wind load (pa) acting on the straight members, only the vertical distance between the intermediate rings is used instead of the projected length in elevation. The wind pressure acting on the intermediate rings themselves is completely overlooked in most of the calculations.
- The bending stresses due to the bending moments caused in the lattice members by the wind loads are only taken into account in the calculations for the Adziogol lighthouse. Although here too, they were of little consequence, because the bending moment used was well below the determinant hogging moment value of a continuous beam.
- The intermediate ring concept is not completely coherent. It is argued that the intermediate rings have to keep the verticals in position and prevent their deflection in bending. In addition, according to the explanations of the way the intermediate rings worked, they would transfer the wind loads to the sides into the intersection points of the verticals. In reality, the heights of the intermediate rings – because of their equidistant arrangement – and the intersection points only seldom coincide with one another. As a consequence the vertical members, which carry lateral loads be-

tween two intersection points, are obviously loaded in bending.

- The intermediate rings and their loads are included only in the calculations for the Adziogol lighthouse. In this case, the bending load and bending deflection of the rings themselves are completely ignored. In addition, the compressive stresses in the rings are not given a buckling check. The stresses in some cases are even above the permissible material stresses given in the calculations.
- As a rule the choice of intermediate ring appears to have been made on the basis of experience; the leg length of the cross section is mostly one half to three-quarters of the leg length of the vertical members.
- In addition to the intermediate rings, other important elements of the structure not given attention in calculations include the main rings of the NiGRES tower, which separate the segments.

The evaluated calculations in general show a high level of consistency. The differences are limited to the assumptions about wind loads, the characteristic values of the materials used and other project-specific particularities. While the earlier calculations at least to some extent are still based on English or Russian units of measurement, from the 1920s onwards the calculations use metric units. An exception to the general similarity of approach is the buckling formulas used in the calculations. These were always being replaced over the period studied in this book by the then current method. The structures of the 1890s used the formulas of Augustus Edward Hough Love, but later the methods came mostly from German-speaking countries, e.g. the buckling formulas published by German engineers Franz Laissle and Adolf Schübler and by the Austrian Johann Emanuel Brik.

Over the whole of the period studied, the use or reference to international literature on the subject is notable. Above all the content of German books and research is used over and over again. In addition to the use of already proven buckling formulas, reference is made in many places to the engineer's handbook published by the Akademischer Verein Hütte.

It is striking that the simplified calculations of the towers stand in obvious contradiction to other structural calculations by Shukhov, which received a lot of attention in the specialist literature and excel though their high scientific standard and precision. The difference to the papers published by Shukhov on arch girders [47] or to his investigations of beams on elastic foundations [48] is remarkable. But also the structural calculations for other building projects are considerably more detailed, such as those for the design of the trusses in the frame for the factory in Lysva. [49]

Despite this and all the discussed inaccuracies and shortcomings, the conclusion remains that the Shukhov calculation model fulfilled its task. With the exception of the collapse in Dnipropetrovsk, in spite of the large number of towers built, no other incidents of damage that could be traced back a failure of the structure are known.

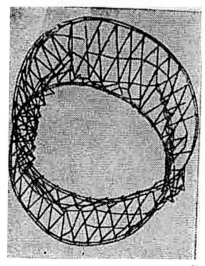

31   Model tests by Gorenšteijn

31

I. число ногъ „n" = 48; β±γ̄ = КРАТНОМУ 7°30

II. число ногъ „n" = 24; β±γ̄ = КРАТНОМУ 15°00'

Примѣчаніе: при числѣ ногъ n=36; β±γ̄ = КРАТНОМУ 10° и при n=30; β±γ̄ = КРАТНОМУ 12°

$$\xi = R\,Cos\,\beta$$
$$\zeta = z\,Cos\,\bar\gamma$$
$$R\,Cos\,\beta = z\,Cos\,\bar\gamma$$
$$\frac{R}{z} = \frac{Cos\,\bar\gamma}{Cos\,\beta}$$
$$\frac{z}{R} = \frac{Cos\,\beta}{Cos\,\bar\gamma}$$

β / γ̄	48°45	50°0	51°15	52°30	53°45	55°0	56°15	57°30	58°45	60°00	61°15	62°30	63°45	65°00	66°15	67°30	68°45	70°00	71°15
	0,6593458	0,6427876	0,6259235	0,6087614	0,5913096	0,5735764	0,5555702	0,5372996	0,5187733	0,5000000	0,4809888	0,4617486	0,4422887	0,4226183	0,4027467	0,3826834	0,3624380	0,3420201	0,3214395
0,9997620	0,6595027	0,6429406	0,6260725	0,6089063	0,5914503	0,5737129	0,5557024	0,5374275	0,5188368	0,5001190	0,4811035	0,4618585	0,4423940	0,4227189	0,4028425	0,3827745	0,3625243	0,3421015	0,3215160
2°30 0,9990482	0,6599759	0,6434000	0,6265198	0,6093414	0,5918729	0,5741228	0,5560995	0,5378115	0,5192675	0,5004763	0,4814470	0,4621885	0,4427101	0,4230209	0,4031304	0,3830479	0,3627833	0,3423455	
0,9978589	0,6607605	0,6441665	0,6272665	0,6100676	0,5925785	0,5748071	0,5567623	0,5384525	0,5198864	0,5010729	0,4820208	0,4627394	0,4432377	0,4235251	0,4036109	0,3835045	0,3632157	0,3427540	0,3221292
5° 0,9961947	0,6618644	0,6452429	0,6283144	0,6110867	0,5935683	0,5757673	0,5576924	0,5393520	0,5207549	0,5019099	0,4828261	0,4635124	0,4439782	0,4242323	0,4042851	0,3841452	0,3638224	0,3433265	0,3226673
0,9940563	0,6632882	0,6466310	0,6296660	0,6124013	0,5948452	0,5770059	0,5588921	0,5405122	0,5218752	0,5029896	0,4838647	0,4645095	0,4449332	0,4251452	0,4051548	0,3849715	0,3646051	0,3440651	0,3233614
													0,4461052	0,4262650	0,4062220	0,3859855	0,3655654	0,3449713	0,3242132
0,9883615	0,6671099	0,6503568	0,6332941	0,6159299	0,5982726	0,5803306	0,5621123	0,5436266	0,5248521	0,5058878	0,4866527	0,4671859	0,4474969	0,4275948	0,4074892	0,3871897	0,3667059	0,3460476	0,3252246
10°0 0,9848078	0,6695172	0,6527036	0,6355793	0,6181525	0,6004315	0,5824247	0,5641407	0,5455883	0,5267762	0,5077133	0,4884088	0,4688718	0,4491117	0,4291378	0,4089597	0,3885869	0,3680292	0,3472965	0,3263982
0,9807853	0,6722631	0,6553805	0,6381860	0,6206877	0,6028940	0,5848154	0,5664544	0,5478259	0,5289566	0,5097955	0,4904119	0,4707948	0,4509536	0,4308379	0,4106369	0,3901806	0,3695386	0,3487206	0,3277368
0,9762960	0,6753544	0,6583942	0,6411206	0,6235418	0,6056665	0,5875025	0,5690592	0,5503450	0,5313688	0,5121397	0,4926670	0,4729596	0,4530272	0,4328793	0,4125252	0,3919748	0,3712378	0,3503242	0,3292439
13°45 0,9713421	0,6787987	0,6617522	0,6443904	0,6267219	0,6087552	0,5904988	0,57/9614	0,5531518	0,5340789	0,5147517	0,4951796	0,4753717	0,4553377	0,4350870	0,4146291	0,3939738	0,3731311	0,3521108	0,3309250
15°0 0,9659258	0,6826050	0,6654627	0,6480037	0,6302362	0,6121687	0,5938100	0,5751686	0,5562535	0,5370736	0,5176381	0,4979562	0,4780373	0,4578909	0,4375267	0,4169541	0,3961830	0,3752234	0,3540853	0,3327778
0,9600499	0,6867828	0,6695356	0,6519697	0,6340935	0,6159155	0,5974443	0,5786889	0,5596580	0,5403607	0,5208062	0,5010039	0,4809631	0,4606934	0,4402045	0,4195060	0,3986078	0,3775199	0,3562524	0,3348154
0,9557170	0,6913432	0,6739815	0,6562990	0,6383040	0,6200053	0,6014115	0,5825315	0,5633742	0,5459489	0,5242645	0,5043307	0,4841568	0,4637525	0,4431276	0,4222916	0,4012546	0,3800268	0,3586180	
0,9469301	0,6962983	0,6788121	0,6610028	0,6428789	0,6244490	0,6057220	0,5867066	0,5674121	0,5478475	0,5280221	0,5079454	0,4876269	0,4670764	0,4463036	0,4253183	0,4041305	0,3827505	0,3611883	0,35945
0,9396926	0,7016611	0,6840403	0,6660939	0,6478303	0,6292585	0,6103873	0,5912255	0,5717823	0,5520670	0,5320889	0,5118576	0,4913826	0,4706738	0,4497410	0,4285941	0,4072432	0,3856985	0,3639702	0,34206
0,9320079	0,7074466	0,6896804	0,6715860	0,6531719	0,6344470	0,6154201	0,5961003	0,5764968	0,5566190	0,5364781	0,5160780	0,4954342	0,4745546	0,4534492	0,4321280	0,4106010	0,3888787	0,3669712	0,34488
22°30 0,9238795	0,7136708	0,6957483	0,6774947	0,6589186	0,6400289	0,6208346	0,6013449	0,5815689	0,5615162	0,5411961	0,5206185	0,4997931	0,4787298	0,4574387	0,4359299	0,4142135	0,3923001	0,3701999	0,34792

ПРИМѢРЪ ПОЛЬЗОВАНІЯ ТАБЛИЦЕЙ: дано R = 25'-7619,9316 ‰ и z = 14'16"= 44/9,5603 ‰; n = 48; z/R = 44/9,5603/7619,9316 = 0,5799999 ближайшее отношеніе по таблицамъ 0,5803306 (β=55°, γ̄ = 8°45) но β±γ̄ не кратно 7

принявъ β = 52°30' найдемъ β = 53°45 и γ̄ =1°15, тогда z/R = 0,59/4523 или β=55° и γ̄ = 2°30, тогда z/R = 0,5741228, второе значеніе является наиболѣе подходящимъ

тогда при данномъ z=44/9,5603 искомое R = 44/9,5603/0,574/228 = 7697,93 ‰, т.е. отличается отъ предложеннаго на незначительную величину = 7697,93−7619,93 = 78 ‰

32 Design tables by Shukhov

Aleksandr Išlinskij writes about Shukhov's approach to the design of structures: "If it is necessary to create an engineering structure, it should not be based on theoretical considerations alone, but also on thoughts of economic viability and engineering feasibility." [50] In the case of hyperbolic lattice structures, the focus was much more on achieving a straightforward design and construction of a complex three-dimensional structure than achieving the greatest possible accuracy in its calculations.

The design process adopted by Vladimir G. Shukhov
Like the structural calculations, the preliminary and detailed design processes for the water towers were very rational and based on standardised methods. Precalculated tables eased the task of the structural engineer in the days before digital calculators and computers, particularly when handling trigonometrical functions.

Use of design tables
The example of a design table (Fig. 32) discussed here comes from the Academy of Sciences in Moscow. It is used to assess very quickly the influence of different design parameters on geometry and form. [51]

This table give the relationship between the angles in the triangle in plan (Fig. 13, p. 31). The rotation angle φ is divided into one part β that includes the angle between the start point of a generating straight line and the (notional) waist point on that line, and another part γ between this waist point and the top end of the line. Equating the expressions for the waist radius R_T thus:

$$R_O \cos\gamma = R_U \cos\beta \rightarrow \frac{R_O}{R_U} = \frac{\cos\beta}{\cos\gamma} = \frac{1}{K_F} \qquad \text{(F 18)}$$

gives the relationship between the top and bottom radii and the two part angles. For the rotation angle φ, which should be a multiple of the central angle ψ, the following applies:

$$\varphi = k \cdot \psi = k \cdot \frac{360°}{2n} = \beta \pm \gamma \qquad \text{(F 19)}$$

The top row of the table contains the angle β and the left hand column contains the angle γ each in steps of 1.25°. The values in the table give the quotient of $\cos\beta/\cos\gamma$, which equals the reciprocal of the K_F-value.
The design table shows possible angle combinations for selected radii and number of members. Under the table is a handwritten calculation explaining the process: For a tower, it gives the bottom radius R_U = 25 feet, the top radius R_O = 14.5 feet and 48 members. The central angle ψ is given as 7.5°. The quotient of RU/RO = 0.5799999. Using this value, the engineer looks for an angle combination for which the value of $\cos\beta/\cos\gamma$ in the table comes the closest to the quotient of R_U/R_O and which is a multiple of the central angle. In the example, 0.5741228 is selected, which is based on β = 55° and γ = 2.5° and therefore gives a rotation angle $\varphi = \beta - \gamma$ = 52.5° – a multiple of 7.5°. The bottom radius is now adjusted using this value to 25 feet 4 inches.

Through the introduction of the waist angle γ the engineer can similarly determine the position of the waist: if the angle γ is subtracted, the notional waist lies above the top ring and the tower appears slightly like a truncated cone in elevation. If γ is added, the tower has a waist. Because of the interaction of the geometry and structural behaviour, the experienced engineer is not only able to determine the geometry but also anticipate and control the flow of force. In addition, the plan view can be used with the geometric relationships to determine the true lengths of the members in three-dimensional space. This allows the engineer to calculate the quantity of steel and quickly arrive at the cost of the proposed tower. A comparison of geometries of built towers shows this table was not used for all the towers. Possibly there were other tables with different angular increments and a larger range of β and γ angles.
Shukhov's design processes can be seen as an early example of parametrised design: the generation of a shape through the variation of few basic parameters. The many actual designs based on a generated parametrised shape show Shukhov's methods to be strikingly similar to contemporary techniques.

Hypothetical flow of the design process for water towers
The design process can be reconstructed from what the structural engineering calculations reveal and the analyses of the built water towers (Fig. 33). Two parameters are known at the start of the project: the desired tank volume V and the height of the tower shaft above ground H required to provide the necessary water pressure.
In a first step, the top ring diameter can be derived from standardised tank size and type. For the Intze tank type, this comes from the tank shape and the standardised dimensions, with flat-bottomed tanks from the geometry of the support grillage on the top ring. In the second step, the number and cross sections of the vertical members could be estimated from the tank load. The wind load acting on the whole system (tank and supporting lattice structure) is then calculated in a third step. From the resulting overturning moment, the diameter of the bottom ring required to provide stability against overturning can be derived. According to the investigated design material, the resisting moment is calculated by taking the empty water tank into account and neglecting the anchor bolts. The calculations aims to achieve a theoretical factor of safety of about 1.05. After the radii, height and number of members, a suitable rotation angle φ has to be found. A favourable angle combination is then selected from the tables as discussed earlier. In spite of having analysed the built examples, it is not possible to say exactly how the choice of rotation angle is made. There are a few clear pointers however:

- A larger rotation angle is usually chosen for high towers, probably on the grounds that the first intersection points of the generatrices would otherwise be too far apart and the tower would be weaker as a result.
- In lower towers, especially in combination with a heavy applied load, the rotation angle is usually smaller. This is due to the fact that the torsional stiffness of larger cross sections would have precluded a more pronounced rotation angle.

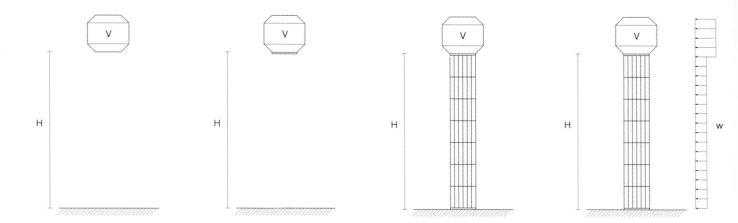

How difficult it is to reconstruct the method of choosing the rotation angle is shown by the water towers in Yefremov (Fig. 34) and Yaroslavl (Fig. 35). Both have the flat-bottomed tanks constructed in the same way with a capacity of 10,000 buckets (123 m³), a lattice structure height of 15.20 m and a top ring diameter of 5.03 m, but the numbers of members are different. The water tower in Yefremov uses 24 vertical members (Σ120/120/12 mm), the one in Yaroslavl 32 (Σ100/100/10 mm). The bottom ring diameters are almost the same: 9.45 m in the case of Yefremov and 9.75 m for the tower in Yaroslavl. The slightly larger diameter of the latter is due to the higher overturning moment caused by the greater number of members. The rotation angles on the other hand differ considerably: 85° for the tower in Yefremov and 56.25° for the tower in Yaroslavl.

In spite of the highly standardised design steps and the hypothetical design process discussed here, it can be concluded that the approach was not automated. In addition to the numerous examples that provide evidence in support of the above described process, there are also some towers that deviate from it. The extent to which, for example, aesthetic criteria such as the subjective sense of proportion played a role in the development of the geometry, cannot be quantified today.

Construction sequence

The foundations are either pads or strips, depending on the number of members. Masonry pad foundations are often used up to 24 members. In this case, the bottom ends of the members are brought together into a common support point or to two support points spaced slightly apart. Where the number of members is greater and strip foundations are used, the support points are generally constructed on alternate sides of the bottom ring (Fig. 38, p. 93). For higher towers, this has the advantage from the point of view of design and construction that the position of the first intersection point is quite low.

For flat-bottomed tanks, spreader beams on the top of the hyperbolic structure take the loads from the radial substructure of the tank (see Fig. 36, p. 92), as is the case on the tower in Tsaritsyn (today Volgograd). The same is not necessary for Intze tanks, which rest on the top ring of the tower.

Evaluation of the water towers built between 1896 and 1930

The development of the water towers built by Shukhov between 1896 and 1930 was determined on the basis of old drawings, photographs and tables. In the archive of the Academy of Sciences, there are two tables that provide the most important key data for towers. The sheets of Opus 1508–84 (1896–1914) and Opus 1508–79 (ca. 1915–1920) contain precise information about the dimensions and weights of the water towers built between 1896 and 1920.

The two tables were translated, combined and the Russian units of dimensions and weight converted to the metric system (see "Towers in comparison", p. 114ff.). In addition, further towers whose geometry and tank volumes were known from other literature, e.g. from the book about water towers by Dmitrij Petrov [52], were included in the tables. The collated material still has some gaps, as only one tower built after 1920 could be included in this evaluation.

The tables contain the most important geometrical parameters, such as the number of lattice members and their cross sections, the height and top and bottom diameters of the lattice structure as well as the precise information about the tank type and volume, the weights and dimensions of the individual elements of the structures and the dimensions of the foundations. Using these values, the construction costs can be easily compared and used for future projects.

On the basis of these examples, the evaluation seeks to show the development of the water towers between 1896 (first tower for the All-Russia Exhibition in Nizhny Novgorod) and 1930 (water tower in Dnipropetrovsk). The newly compiled table covers a total of 39 towers. However, precise geometrical information could not be reconstructed for all the towers, because the original tables did not contain the required rotation angle φ nor the number and arrangement of the intermediate rings. Additional photographs and

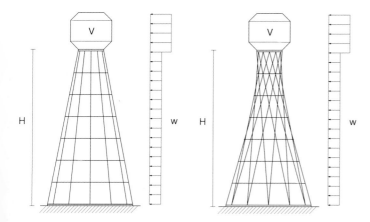

33

33 Schematic reconstruction of the design process for a water tower
34 Water tower in Yefremov (RUS) 1902
35 Water tower in Yaroslavl (RUS) 1904

drawings were necessary to obtain this information. By counting the number of intersection points of the vertical members, it was thus possible to reconstruct the rotation angle of 18 towers. The following dependencies and developments can be gleaned from the collated material:

- The achieved tower heights (just the lattice) varied between 8.53 m (location unknown) and 43.50 m (Dnipropetrovsk). The average height is around 20 m, with a slightly increasing tendency for greater heights in the later years.

- The volumes of the installed water tanks vary greatly between 6 m³ (Sagiri) and 1230 m³ (Voronezh); the most frequently used tank size is 123 m³, which is equal to the Russian volume of 10,000 buckets. Of the 39 investigated towers, 24 have a flat-bottomed tank, nine an Intze tank and six a suspended bottom tank. No relationship between time and tank size can be established; however, it is clear that Intze tanks are always used for large capacities (> 400 m³). The tank sizes are standardised in most cases and are 10,000, 15,000, 30,000, 50,000 and 100,000 buckets (1 bucket = 12.3 l).

- The top radii largely depend on the tank size (independent of the tank type). The tanks were available only to certain standardised patterns depending on the particular manufacturer and therefore the same top radii come up again and again. Therefore the diameter 5.79 m is used for flat-bottomed tanks with volumes of 123 m³, 5.03 m for 184 m³ and 6.10 m for 369 m³. Although there are always deviations from this pattern, which were probably due to variations in the design of the tanks.

- On the other hand, there were no recurring values for the bottom radii because, as mentioned earlier, this value arises from the calculation of the required restoring moment to resist overturning.

- Correspondingly the resultant K_F-values also varied significantly: from 1.61 (two water towers in Tambov, 738 m³) to 3.92 (Moscow Simonovo, 28.3 m³).

- The horizontal intermediate rings are normally equally spaced vertically over the height. In some cases, the vertical spacings

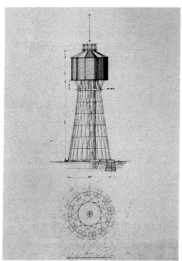

34

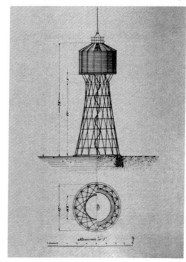

35

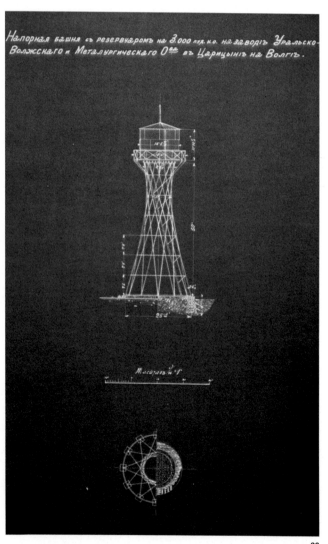

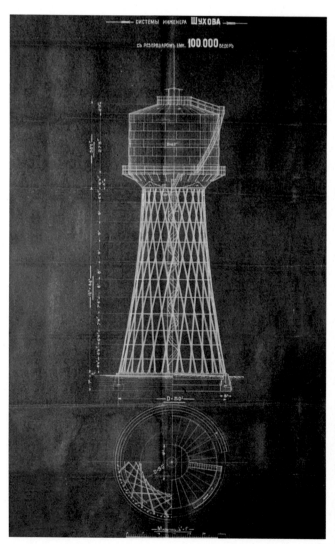

36

37

38

of the rings also increase with height, particularly when there are more intersection points of the vertical members in the upper part of the tower and this area is therefore stiff enough as it is.

- The vertical distances of the intermediate rings fluctuate between 1.80 and 3.20 m. On average the distance is about 2.40 m.
- High rotation angles between 90 to 105° were only used with towers with very small tank volumes, between 6 m³ (Sagiri) and 37 m³ (Tsaritsyn). The number of members varies in these cases between 24 and 40, the leg lengths of the angle profiles are a maximum of 75 mm. One exception is the first water tower in Nizhny Novgorod, which with a rotation angle of 94.5° and a tank volume of 117 m³, values which are above those of the later towers.
- For water towers with large tank volumes (> 500 m³), the rotation angle is normally under 75°; the cross sections are generally double U-profiles between 100 and 140 mm deep. One exception is the water tower at Mykolaiv, which with a tank volume of 615 m³ has a rotation angle of 82.5°. Here angle sections were used.

In spite of the above-mentioned increase in achieved height and installed tank sizes for some towers in the later years, the analysis of the examples shows no clear line of development over the time period studied.

In addition to the translation and evaluation of the two tables, the chapter "Towers in comparison" (p. 114ff.) also contains two graphs that show the height, rotation angle and number of members of individual towers and the installed tank volume (p. 120f.). In addition, the chapter contains drawings of 18 water towers, whose precise geometries have been deduced (see p. 122ff.).

36 Water tower in Tsaritsyn (RUS) 1899
37 Water tower in Voronezh (RUS) 1915
38 Support detail of the water tower in Nizhny Novgorod (RUS) 1896, vertical members on alternate sides

NiGRES tower on the Oka

Telescopic construction method   96
Geometry of the sections   98
Constructional details   101
Results of new calculations   103
Summary   110

NiGRES tower on the Oka

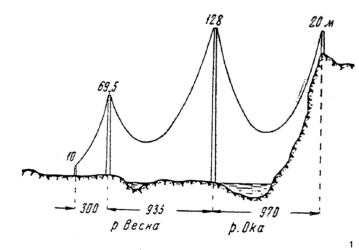

1

a

b

2

Without doubt, the NiGRES high-voltage electricity masts on the Oka are the most impressive lattice towers ever built by Shukhov. The ensemble, which once comprised two hyperboloid towers each with three sections and two towers each with five sections, lies 35 km south-west of Nizhny Novgorod near the city of Dzerzhinsk. Their great height was necessary for the conductors to span a clear 970 m across the Oka (Fig. 1). Shukhov began the design in 1927; the whole ensemble was completed in 1929. The 110 kV line operated until 1989, when it was decommissioned and the two smaller approach masts disassembled. During the following years, the river undermined the foundation of the tall mast closest to the river with the result that it collapsed. Since then only one tower, which has a height of 130.2 m, remains (Fig. 3).

In summer 2005, some of the loadbearing steel members of the last remaining mast were removed without permission. The perpetrators, who were never identified, severed 16 of the 40 vertical members in the lower segment (Fig. 2 a) and took off the two lower horizontal rings with cutting torches. [1] Without these members, the tower was in acute danger of collapse. An international team of construction researchers and engineers succeeded in raising the awareness of the responsible authorities of the danger to the tower and subsequently undertook its refurbishment using funds provided by the local electricity supply company. [2] In the sense of a reconstruction rather than a strengthening project, the missing elements were replaced in March 2008 with new members of almost the same cross section (Fig. 2 b). High-strength friction-grip bolts were used instead of rivets. The red anti-corrosion paint differentiates the new members and makes clear where the repair work has been carried out (Fig. 20 a, p. 107).

Telescopic construction method

The two smaller approach masts were built using conventional erection methods, while the two taller towers used the telescopic method that had already been employed to build the Shabolovka radio tower in 1922. With this method, after the first section has

---

1 Schematic drawing of the NiGRES
   transmission masts
2 NiGRES tower on the Oka,
   Dzerzhinsk (RUS) 1929
   a Gaps in the loadbearing
     structure
   b Repairs carried out in 2008
3 NiGRES tower on the Oka,
   condition in 2007

3

been completed the following segments are preassembled inside the shaft and pulled upwards one by one using A-frames and pulleys. A type of wooden corset holds the vertical member bottom ends together at every section. [3] When the new section clears the top of the section already in place, the corset is released and the lattice members connected with those of the lower segment (Fig. 4).

The use of the telescopic erection method also affected the form of the hyperboloid sections. Large rotation angles and the resulting pronounced necking of the sections had to be avoided, otherwise this would have obstructed the passage of the next section through the previous one.

Geometry of the sections

In elevation, the geometry of the tower approaches that of a cone. The ring diameters of the segments decrease from 34.0 m at the foundation ring in steps through 25.8 m, 19.4 m, 14.0 m, 10.0 m to 6.0 m at the top end of the shaft, just below the cross arm. The lower four sections are each 24.9 m high, the fifth and top section is 24.3 m high. The conductors, which have been removed, were attached at a height of 128.0 m; the overall height of the structure including the cross arms is 130.2 m (Fig. 5). The lower three segments each have 40 vertical members; the two upper rings have 20. This results in central angles of 9° and 18° for the lower and upper segments respectively. The rotation angle for the lower two sections is 36° and this increases to 40.5°, 54° and finally 72° at the top. The top section is the only one to have noticeable necking. In three lower sections and in the top, the fifth, section the two families of vertical members cross one another four times. The fourth section has only three family crossings. The rotation angles and geometric boundary conditions cause the inclination of the members to the vertical (the angle ε see "Geometry of the frame mesh", p. 31) to continuously decrease from bottom to top. Thus this angle is 21.94° in the first section and only 11.57° in the top section – the members become steeper from bottom to top. The precise dimensions and angles are summarised in Fig. 7 (p. 100).

4   Erection of one of the two tall NiGRES towers using the telescopic method
5   Individual tower sections

4

130.2 m

34.0 m   25.8 m   19.4 m   14.0 m   6.0 m
                                    10.0 m

24.3 m
24.9 m
24.9 m
24.9 m
24.9 m

φ = 36°   φ = 36°   φ = 40.5°   φ = 54°   φ = 72°

5

5th section:

4 intersection points
φ = 72°
9 intermediate rings
20 vertical members

4th section:

3 intersection points
φ = 54°
10 intermediate rings
20 vertical members

3rd section:

4 intersection points
φ = 40.5°
10 intermediate rings
40 vertical members

The ends of the vertical members coincide at the top ring

2nd section:

4 intersection points
φ = 36°
10 intermediate rings
40 vertical members

1st section:

4 intersection points
φ = 36°
10 intermediate rings
40 vertical members

	1st section	2nd section	3rd section	4th section	5th section
$2n$ [–]	40	40	40	20	20
R_U [m]	17	12.9	9.7	7	5
R_O [m]	12.9	9.7	7	5	3
H [m]	24.9	24.9	24.9	24.9	24.3
φ [°]	36	36	40.5	54	72
k intermediate rings [–]	10	10	10	10	9
K_F-value (R_U/R_O) [–]	1.32	1.33	1.39	1.40	1.67
Ψ [°]	9	9	9	18	18
n_{SP} [–]	4	4	4	3	4
G [m]	10.0	7.62	6.31	5.73	4.97
L [m]	26.84	26.04	25.69	25.55	24.80
ε [°]	21.94	17.01	14.22	12.96	11.57
R_T [m]	12.85	9.65	6.99	4.94	2.87
γ [°]	4.88	5.55	3.42	8.89	17.01
β [°]	40.88	41.55	43.92	45.11	54.99

7

	Vertical members [mm] (number)	Intermediate rings [mm]	Main rings (spacing) [mm]
1st section	L120/120/12 (40)	L80/80/10	L100/100/12 (240)
2nd section	L100/100/12 (40)	L75/75/8	L90/90/9 (200)
3rd section	L100/100/10 (40)	L75/75/8	L75/75/8 (200)
4th section	L100/100/12 (20)	L60/60/6	L75/75/8 (200)
5th section	L100/100/10 (20)	L50/50/6	L75/75/8 (10)

8

6

Structural arrangement

The circular foundation of only lightly reinforced concrete has a truncated cone cross section almost 3 m deep. Two angle profiles run 53 mm apart along the top. The distance between the angle profiles is set so that the holding down bolts (45 mm diameter) can fit between them. At the 40 support points, a slightly angled connection plate (12 mm thick) with additional packing plates is introduced between the angles (Fig. 9). The vertical members come down and connect to these angled plates on the inside or the outside. The distance between the two vertical members at the supports created by the thickness of the connecting plate is continued in the lattice at the intersection points: an almost square plate with a thickness of 12 mm sits between the legs of the vertical members at the intersection points. The connection is made with two rivets.

Only equal-legged steel angle profiles are used to construct the tower, with the exception of the cross arms. The cross section of the straight members decreases from Σ 120/120/12 mm at the bottom to Σ 100/100/10 mm at the top (Fig. 8). The first three sections consist of 40 straight members, the upper two 20. At the top end of the third segment therefore, thanks to an increase in the rotation angle of one half of the central angle, the ends of the members meet in pairs to create triangles. These form the support for the lattice members in the fourth section (Fig. 6). The junction of the hyperbolic sections is stiffened by a light horizontal truss. All horizontal intermediate rings are positioned at equal vertical distances from one another. The intermediate rings are connected to the verticals by a small cantilever bracket. These brackets are made up of a triangular plate and an angle profile "arm" and are connected to the horizontal intermediate ring with a bolt. These are the only bolted connections in the whole of the structure; all other connections are riveted (Fig. 10).

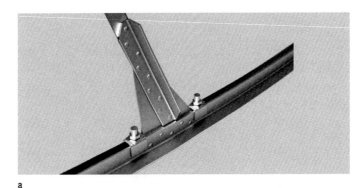

a

b   9

a

b   10

6   Individual tower sections, geometry, rotation angle and intersection points
7   Overview of dimensions and geometric relationships
8   Summary of member cross sections
9   Details of the support points
    a Rendering
    b Exploded view
10   Connections of the members to the intermediate ring
    a Rendering
    b Condition 2011

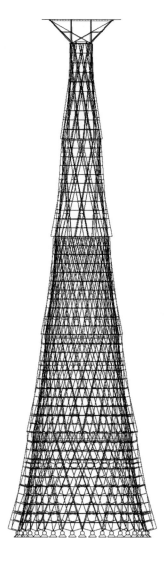

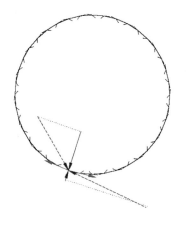

11

NiGRES tower on the Oka

Results of new calculations

The original calculations by Shukhov have been discussed in the section "Structural calculations for the NiGRES tower on the Oka" (p. 79ff.). In the course of the refurbishment project, the design of the NiGRES tower was accurately modelled using finite-element analysis software and the structural behaviour under self-weight and wind loads investigated. The results of these calculations are compared at the end of this chapter (p. 110) with Shukhov's original design.

Structural behaviour under self-weight

As anticipated with self-weight alone acting on the tower, all the straight members are in compression. Because the inclination angle ε with respect to the vertical of the lattice members decreases segment by segment with height up the tower, the compressive force in the members increases stepwise just below each ring beam. The ring beams themselves are in tension.

This appears to contradict the observations made in the section "Vertical load transfer" (p. 32f.), where the top ring of hyperboloids without a waist must be designed as a compression ring to keep the lattice members in equilibrium. This is valid but applies only to individual segments. Where one segment is placed on top of another, the influence of the upper segment is much greater. Its members meet the rings – when viewed in plan – at more of an angle, which means the normal force component is relatively high. The member ends of the lower segment on the other hand generally meet the ring much more tangentially, therefore the component acting perpendicular to the ring is correspondingly smaller (Fig. 11). Overall the loads from the members above create tensile forces in the ring, which fluctuate due to superimposed smaller tangential components of the members.

The internal forces of the vertical members and main rings under the action of self-weight (load case 1) are shown in the table in Fig. 18 (p. 106).

Structural behaviour under horizontal wind load

The reference wind speed with a 0.02 per cent probability of being exceeded in any one year (50-year event) was established for Dserschinsk in consultation with the university in Nizhny Novgorod as $v_{ref} = 22.5$ m/s. Using this reference value, the wind loads can be determined in accordance with DIN 1055-4. [4] It was demonstrated that the structure is not sensitive to vibrations from the action of gusts in accordance with Section 6.2 of the standard and therefore the resonance effects caused by gusts are ignored.

Determination of wind loads

The associated dynamic pressure is therefore $q_{ref} = 0.32$ kN/m². The wind speed and dynamic pressure equate to wind zone 1 in DIN 1055-4. The total wind force acting on a component of the structure is calculated from

$$F_W = c_f \cdot q(z_e) \cdot A_{ref} \qquad \text{(F01)}$$

using the force coefficient c_f, the reference height z_e and the reference area A_{ref}. The reference height is equal to the height above

12

11 Normal force diagram under self-weight (compressive force red), typical force resolution at the first main ring: the resultants of the normal force components acting on the ring create a tensile force (blue).
12 View looking down from the top of the tower
13 Table of horizontal wind loads in accordance with DIN 1055-4
(H = height of segment top edge; z_e = height above ground; $q(z_e)$ = dynamic wind pressure at height above ground z_e)

	H_u [m]	H_o [m]	z_e [m]	$q(z_e)$ [kN/m²]	$q(z_e)$ 1.84 [kN/m²]
1st section	0	24.9	15	0.632	1.16
2nd section	24.9	49.8	40	0.909	1.67
3rd section	49.8	74.7	65	1.053	1.94
4th section	74.7	99.6	90	1.139	2.10
5th section	99.6	123.9	114	1.205	2.22
Cross arm	123.9	130.2	127	1.237	2.28

13

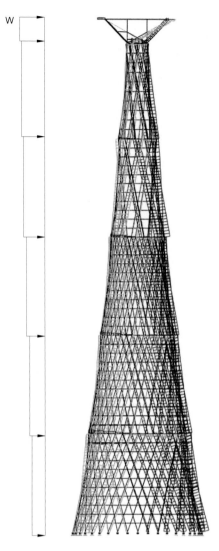

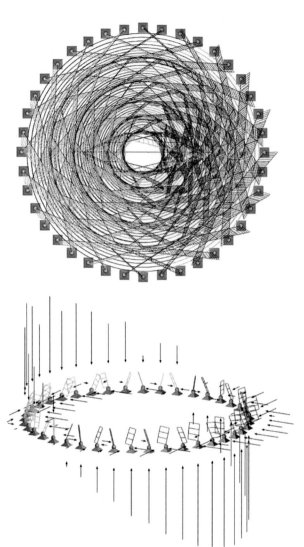

14

14 Normal force diagram for load case combination 1 (self-weight and horizontal wind load in accordance with DIN 1055-4)

15 Schematic representation of the vertical member forces on the main rings: resolution into normal and tangential force components

16 Loads and deformations at the first main ring under load case 2 (horizontal wind load in accordance with DIN 1055-4)

a Forces on the main rings from the vertical members
b Normal and tangential force components
c Resulting normal force distribution
d Vertical bending moments
e Horizontal bending moments
f Deformed shape

ground of the bottom of the section plus 0.6 times the section height. The force coefficients for components with an angular, non-rounded types of cross section are calculated from the following formula:

$$c_f = c_{f,0} \cdot \psi_\lambda \qquad \text{(F 02)}$$

The force coefficient for angle profiles is given as $c_{f,0} = 2.0$ for all wind flow directions. The effective slenderness for polygonal cross sections is calculated in accordance with Table 16 of DIN 1055-4 as $\lambda = 70$, the reduction factor using the effective slenderness in accordance with Fig. 26 (DIN 1055-4) as $\psi_\lambda = 0.92$. The critical force coefficient is then calculated from the standard in accordance with equation F 02 as $c_f = 2.0 \cdot 0.92 = 1.84$. The gust dynamic pressure, which depends on height, is determined in accordance with Section 10.3 for non-coastal regions for mixed terrain categories II and III as:

$$q(z) = 1.7 \cdot q_{ref}\left(\frac{z}{10}\right)^{0,37} \text{ für } 7 \text{ m} < z \le 50 \text{ m} \qquad \text{(F 03)}$$

$$q(z) = 2.4 \cdot q_{ref}\left(\frac{z}{10}\right)^{0,24} \text{ für } 50 \text{ m} < z \le 300 \text{ m} \qquad \text{(F 04)}$$

Taking into account the cross-sectional width, the wind loads were applied as uniformly distributed loads on the members. The current wind standards contain no shielding factors for cylindrical or other complex three-dimensional lattice structures. Therefore the full wind loads have to be applied in the calculations, even if it is a conservative assumption that would probably not correspond with reality. In areas on the sides of the structure, the actual wind loads will very probably be much less, due to the close effective spacing of the vertical members. For this reason, wind tunnel tests are being carried out as part of a current research project from which some more accurate statements on the applied wind forces are expected to emerge. The results of the following calculations are therefore to be considered mainly as qualitative and indicative of the principal load transfer behaviour under vertical-horizontal loading, and help to discern the most heavily loaded elements of the structure.

Flow of forces

The global behaviour of the tower appears at first glance to be that of a shear-stiff lattice tube, in which the lattice members on the leeward face are in compression and those on the windward face in tension (Fig. 14). However, on more detailed consideration of the vertical bearing forces, it can be seen that the values do not vary linearly but discontinuously over the cross section. The reasons for this are discussed in the section "Horizontal load transfer" (p. 34ff.). The slope and direction of the the abutting vertical members change at the main rings. Because the inclination of the members relative to the vertical increases towards the bottom but the vertical force component remains unchanged, the member forces increase stepwise just below each main ring ($\Sigma V = 0$). The normal and tangential forces acting on the ring can be calculated by adding the horizontal force components of the vertical members at each transition (Fig. 15).

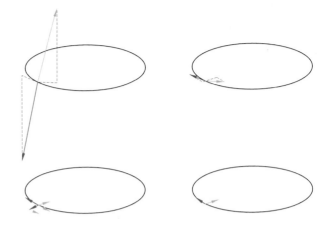

15

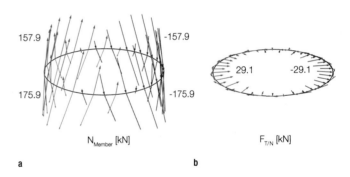

157.9   −157.9

175.9   −175.9

N_{Member} [kN]

a

29.1   −29.1

$F_{T/N}$ [kN]

b

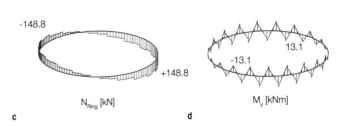

−148.8

+148.8

N_{Ring} [kN]

c

13.1

−13.1

M_y [kNm]

d

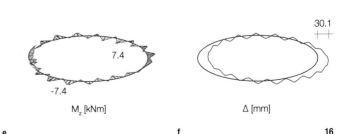

7.4

−7.4

M_z [kNm]

e

30.1

Δ [mm]

f

16

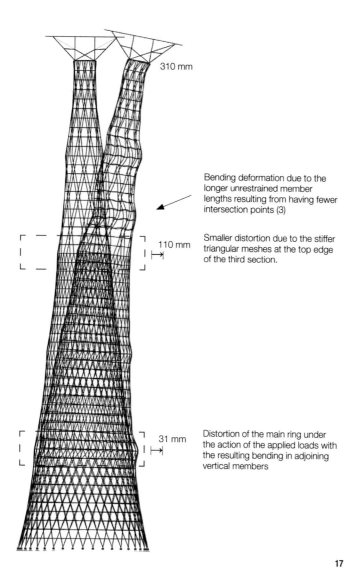

310 mm

Bending deformation due to the longer unrestrained member lengths resulting from having fewer intersection points (3)

110 mm

Smaller distortion due to the stiffer triangular meshes at the top edge of the third section.

31 mm

Distortion of the main ring under the action of the applied loads with the resulting bending in adjoining vertical members

17

	LC 1	LCC 1	LCC 2	LCC 3
1st section members:				
max. N [kN]	-29.5	-262.8/201.0	-389.5/306.0	-153.8/101.9
max. σ_V/σ_{VRd}	0.06	1.04	1.63	1.03
1st main ring:				
max. N [kN]	19.8	-159.1/202.7	-242.0/300.8	-50.6/118.5
max. M_y [kNm]	0.2	20.83	31.23	10.04
max M_z [kNm]	0.2	6.59	10.95	7.49
max. σ_V/σ_{VRd}	0.04	1.48	2.21	0.74
2nd section members:				
max. N [kN]	-18.8	-220.1/179.3	-326.4/272.0	-121.2/82.4
max. σ_V/σ_{VRd}	0.06	1.28	1.92	0.90
2nd main ring:				
max. N [kN]	10.6	-104.9/128.4	-157.7/190.8	-38.9/68.9
max. M_y [kNm]	0.08	12.88	19.32	4.48
max. M_z [kNm]	0.04	7.18	10.53	4.85
max. σ_V/σ_{VRd}	0.03	1.29	1.93	0.52
3rd section members:				
max. N [kN]	-11.80	-171.8/146.2	-256.7/221.2	-92.3/61.1
max. σ_V/σ_{VRd}	0.05	1.02	1.60	0.41
3rd main ring:				
max. N [kN]	6.4	-103.3/120.0	-156.2/178.4	-32.1/50.7
max. M_y [kNm]	0.08	1.05	1.56	0.86
max. M_z [kNm]	0.13	4.93	7.37	5.82
max. σ_V/σ_{VRd}	0.04	0.33	0.46	0.20
4th section members:				
max. N [kN]	-12.70	-218.8/191.7	-332.6/289.9	-96.1/61.6
max. σ_V/σ_{VRd}	0.03	1.75	3.13	0.37
4th main ring:				
max. N [kN]	4.4	-48.6/62.3	-73.9/92.4	-12.7/24.6
max. M_y [kNm]	0.1	16.41	24.61	2.98
max. M_z [kNm]	0.22	8.62	14.36	3.25
max. σ_V/σ_{VRd}	0.02	2.00	4.00	0.65
5th section members:				
max. N [kN]	-7.07	-147.8/130.3	-220.4/196.7	-53.4/38.1
max. σ_V/σ_{VRd}	0.03	1.45	2.95	0.33
Vertical support force [kN]	27.4	245.8 (-188.1)	364.4 (-286.4)	142.7 (-73.9)

18

a

b

19

The unequal horizontal force components acting on the main ring lead to a distribution of normal force which increases from the neutral axis to the edges of the tower cross section. The forces on the windward side are compressive while the forces on the leeward side are tensile. The force components acting on the ring also give rise to bending moments; as a result the light truss of the main ring deforms (Fig. 16, p. 105). In addition to these loads in the horizontal plane, the horizontal shear (in combination with the alternately acting tensile and compressive forces) causes a vertical distortion of the rings. This distortion is dependent on the arrangement of the vertical members below the ring. If these are connected individually to the ring – as is normally the case – the meshes here are an open trapezoidal shape and become heavily deformed (Fig. 19 a). The bending moments about the y-axis are in this case about 1.5- to 3-times larger than the moments about the z-axis (Fig. 18). On the other hand if the meshes here are triangular, which they are when the member ends meet in pairs at the ring, then the connection is considerably stiffer. The vertical distortion and bending moments (about the y-axis) are small in this case.

The deformation of the main rings leads to a characteristic deformed shape of the whole tower. The horizontal and vertical distortion of the ring trusses also affects the adjoining vertical members and intermediate rings, which deform similarly because of their low flexural stiffness. Therefore lateral outward bulges, which can be clearly seen in Fig. 17, form at the transitions between the hyperboloid sections. In elevation, it is noticeable that the bulging is absent at the transition between the third and fourth sections (Fig. 19 b). This can be explained by the above-mentioned arrangement of the members. The top ends of the 40 members of the third section meet in pairs and form the support for the 20 members of the fourth section. The member ends come together in pairs to create stiff triangular meshes, which prevent the large horizontal deformations in this area, in contrast to the otherwise weaker trapezoidal meshes. A further peculiarity of the fourth section is the bulging of the middle of the hyperboloid. This can be explained by the reduced number of intersection

a

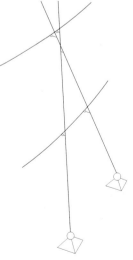

17 Deformed shape under horizontal wind load, 50≈ magnification (max. deformation: 310 mm)
18 Table of vertical member forces and loads in each tower section: Normal force N (compression/tension), bending moment M, stress ratio σ_v/σ_{vRd}
19 Distortion of two meshes with the adjoining main ring under horizontal wind load, 50≈ magnification
    a First main ring with trapezoidal meshes (transition from first to second segment)
    b Third main ring with triangular meshes (transition from third to fourth segment)
20 Modelling of the node points and bearings
    a Actual condition
    b Drawing of finite-element model

b

20

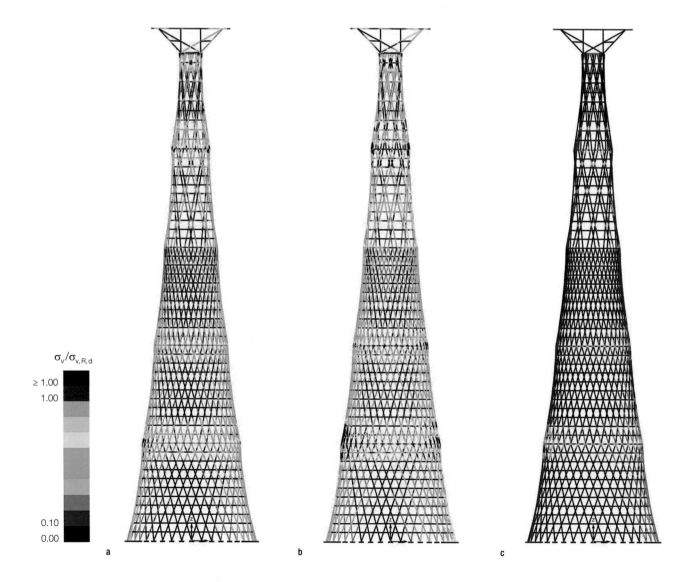

$\sigma_v/\sigma_{v,R,d}$

≥ 1.00
1.00

0.10
0.00

a   b   c   21

points of the members in this segment. The correspondingly longer unrestrained member lengths between the three intersection points means the bending deformations are greater. These deformations arise mainly from the wind load forces transferred from the intermediate rings. In spite of the deformations described here, the tower as a whole is a relatively stiff structure. The maximum movement of the top end of the cross arm is 310 mm, which equates to a deflection ratio of H/422.

Results of the analysis

The calculations were carried out as described in the parametric studies in the chapter "Relationships between form and structural behaviour" (p. 50ff.) assuming grade S 235 structural steel, as confirmed by the staff of the University of Nizhny Novgorod. The structure is assumed to have pinned supports (Fig. 20 b, p. 107). The vertical members are arranged in two layers and connected by flexurally stiff joints at their "intersection points", while the joints between the intermediate rings and the cantilever brackets on the vertical members are pinned. Using the method explained in the section "Choice and size of imperfection shape" (p. 43), the imperfections were applied to the system. The critical buckling load of the whole system was calculated as P_{crit} = 163 kN.

In the course of the calculations, in addition to the load cases (LC) self-weight and wind, three load case combinations (LCC) were investigated:
• Load case 1: Self-weight
• Load case 2: Horizontal wind load in accordance with DIN 1055-4 (Fig. 13, p. 103)
• Load combination 1: LC 1 + LC 2
• Load combination 2: 1.35 LC 1 + 1.5 LC 2
• Load combination 3: 1.35 LC 1 + 1.5 · 0.4 LC 2 (without conductors)

While the load case combinations 1 and 2 represent the original situation with the conductors in place and the wind loads acting upon them, load case combination 3 considers only the current scenario without these effects.
The stresses resulting from the wind loads, which as described above were applied without reduction to the members, were considerably in excess of the permissible stresses, even for load case combination 1 without partial safety factors. This manifests itself particularly strongly in the large deformations on the first and fourth main rings and the adjoining lattice members. If factored loads are used as in LCC 2, the actual stresses exceed the permissible equivalent stress by a multiple of four. The critical internal forces for the load cases or the load case combinations can be found in the table in Fig. 18 (p. 106). Because the loads determined in accordance with the standard would have led to a stability failure of the structure, only 40 % of the wind loads were applied in LCC 4, although the associated partial safety factors were still applied to the load side. In this case, the permissible buckling stresses and equivalent stresses (Fig. 21 c, p. 108), taking into account the 10 % local overstress allowed by DIN 18800 for plastic yield, were not exceeded. [5]

Vertical members	LCC 4	Shukhov's calculations	Difference
1st section	-290.9/248.9	-253.0	15.0 %
2nd section	-235.2/229.5	-194.0	21.2 %
3rd section	-178.7/149.8	-137.0	30.4 %
4th section	-218.5/185.6	-173.0	26.3 %
5th section	-139.2/119.2	-86.0	61.9 %
Vertical support force	269.8/-230.4	253.0	6.6 %

22

21 Stress ratio σ_v/σ_{vRd}
    a LCC 1: Self-weight and wind unfactored
    b LCC 2: Self-weight and wind with safety factors
    c LCC 3: Self-weight and 40 % of the full wind loads in accordance with DIN 1055-4, with safety factors, conductors not present
22 Comparison of the vertical member forces under Shukhov's assumed loads
23 Plan showing the highest loaded and second highest loaded (dashed line) vertical members in each segment on the compression (red) and tension side (blue)

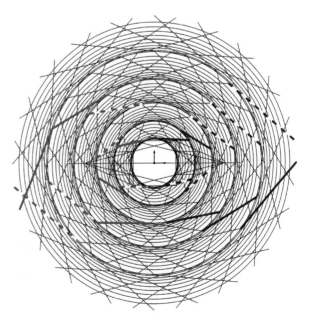

23

Comparison with the original calculations by Shukhov

In addition to the three load case combinations discussed above, a further load case combination LCC 4 considered the load assumptions that Shukhov found were critical in his analysis and design of the tower.

The results of the finite-element analysis deviate substantially from the internal forces and support reactions calculated by Shukhov. There are several reasons for this. Firstly the deviations arise from ignoring the slope of the lattice members; the historical calculations considered only the vertical components of the forces. This process also affects the applied wind loads: because these are applied only to the vertical lengths of the members, their overall value is too small. Taking the first segment as an example, a comparison of the values calculated for the true lengths in Fig. 7 (p. 100) and the vertical lengths of the members shows a difference of 7.8 %. The main reason, however, lies in Shukhov's method of calculation, which is based on the assumption that the most heavily loaded members in each segment are the ones with the greatest lever arm in the direction of the load (see Shukhov's calculations for the lattice towers (p. 70ff.)). Figure 23 (p. 109) shows that this is incorrect. The members with the largest or second-largest compressive or tensile forces in each segment are highlighted in colour. It is noticeable that the start points of some of these members are a long way from the x-axis. The reason for the asymmetric positions of the most heavily loaded members is the angle of 4.5° between the cross arm and the axis of symmetry of the support points in the horizontal projection.

Fig. 22 (p. 109) summarises the internal forces and bearing reactions as well as the differences between the finite-element analysis and the original calculations. As can be seen, the differences increase from the bottom to the top of the tower – with the exception of the fourth section – to reach 61.9 % in the fifth section.

Summary

The outstanding significance of the structure lies above all in its design and detailing. This is revealed in the consistent clarity of the detail and the fascinating lightness of the structure. The innovative construction process with the telescopic method of erection, which must have been precisely thought out during the design of each section, shows the sophistication and artistry of its designer.

Although it has been shown that Shukhov's method of calculation considerably underestimated the forces acting in the structure under the applied loads, the structure has survived the last 80 years almost undamaged. An impressive proof of its structural capability was provided between 2005 and 2008, when many of the members were removed from the first section.

The calculations discussed here show how difficult it is even today precisely to calculate the actual forces in the vertical members under horizontal wind loads. Realistic statements about the vertical member forces and safety factors of the structure are difficult to make without wind tunnel tests to determine the exact effects of the wind. Notwithstanding the above, the analysis of the structural behaviour does reveal a weak point in the design, although it has not led to any serious consequences. The critical

area of the structure appears to be the too lightly constructed trusses separating the five sections from one another. These main rings carry forces from above and below as well as horizontal wind loads while experiencing large bending moments and deformations. The deformations affect the adjoining vertical member ends and neighbouring intermediate rings. By the creation of trapezoidal meshes below the main rings, the bending moments are normally much greater about the y-axis of the beam than they are about its z-axis. An increase in stiffness about the y-axis of the main rings is therefore particularly necessary. An alternative would be to ensure that all the vertical members met in pairs at each main ring beam. In this context, it is worth mentioning that the historical calculations contain no information about how the main rings were analysed or designed. Obviously it can be concluded they were proportioned to provide stiffness to the lattice in these areas.

Consequently, the question arises as to what extent the first design for the Shabolovka radio tower in Moscow, with a height of 350 m and a foundation ring diameter of 90 m, would have been realisable (Fig. 1, p. 9, also Shabolovka radio tower, p. 22f.). The first main ring diameter would have been 72 m – about three times that of the first main ring of the NiGRES tower (25.8 m). It is hardly imaginable that the ring beam would have been stiff enough, at least in the horizontal direction. A horizontal truss would in all probability be too weak; it is conceivable that spoke wheels like the ones at the Adziogol lighthouse (Fig. 17, p. 75, and the structural calculations for the Adziogol lighthouse, p. 75ff.) could have been used to ensure the required stiffness in the horizontal direction.

With its more proportionate form and and unmatched lightness of construction, the NiGRES tower is a further development of the finally constructed version of the Shabolovka tower. After over three decades of experience with these novel structures, the towers on the Oka marked the high point of Shukhov's career.

24 Elevation and plan of the five segments, blueprints of the NiGRES tower from 1928

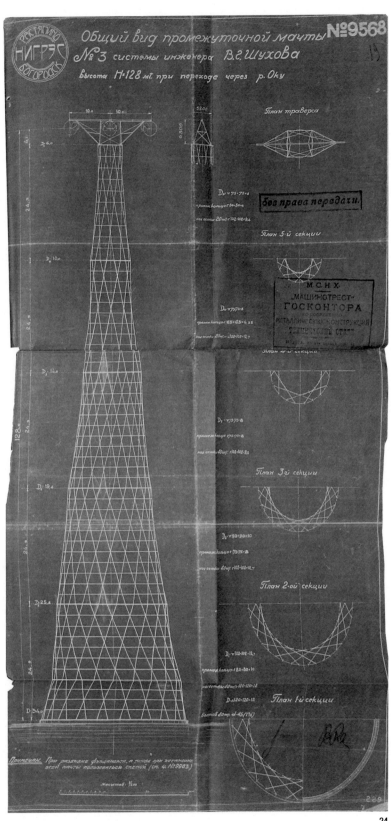

Résumé

1   View from the NiGRES tower across the Oka, Dzerzhinsk (RUS) 1929

At the beginning of the publication, the focus is on the development of hyperbolic lattice towers, their place in the history of construction and the career of Vladimir G. Shukhov. The first part discusses the mathematical principles and geometric relationships of this new type of structure. Following on from this, the main ways the structures transfer vertical and horizontal loads are explored. Using extensive parametric studies and load capacity calculations, the author analyses the interactions of form and structural behaviour. These investigations demonstrate the high efficiency of Shukhov's lattice towers and compare the performance of different mesh variants. The influence of individual shape parameters is discussed and the findings summarised in the chapter "Relationships between form and structural behaviour" (p. 50ff.). The structural analysis of various Shukhov-built water towers often reveals very low reserves of safety. Some of the analysed examples are at the limit of their load-carrying capacity.

The analysis of the original calculations provides an especially informative insight into a very pragmatic approach. Particularly when compared with other structural calculations prepared by Shukhov, for example his arched girders stiffened with ties or his steel tank constructions, which are noted for their precision and technical expertise, the lattice tower calculations feel rather approximate and less detailed. The structural models upon which the calculations are based have to be seen as inadequate in places, even by the standards of the time. Shukhov used the model of a shear-stiff resolved tubular cross section and determined the lattice vertical member forces from the moment of inertia. However, it can be shown that the horizontal loads are mainly taken by the bipods arranged at various angles in space. The amount of load they carry depends on the position of the lattice member pair and the rotation angle. Therefore the resulting lattice vertical member forces are considerably underestimated by the Shukhov calculation model. Furthermore some important elements, such as the intermediate rings, are not designed but are proportioned to suit the practicalities of the structure.

Structural models are simplified representations of reality [1], the results of which must be shown to be sufficiently valid. Even though the model used by Shukhov had its shortcomings, the facts speak for themselves: apart from the collapse of the water tower in Dnipropetrovsk, no other faults that led to a failure of the structure have been found – in spite of the huge number of towers built. The method had therefore proved itself despite inherent deficiencies and remained basically the same, with very little modification, over the whole period these structures were being designed.

The design process for the water towers is reconstructed from the analysis of the structural calculations and the evaluation of the data in tabulated project lists. Even though the process is far from automated, the standardised, table-based design process indicates a very rational approach and a highly schematic work flow. Evidently Shukhov focussed more strongly on the economic aspects of the design and construction processes than on achieving the greatest possible precision in the calculations for these novel three-dimensional structures. Otherwise the considerable number of completed towers would have been much less. Conspicuous in this context is the frequent use of the same elements in any one structure. In contrast to other steel shell structures or trusses of the time, in which each component was designed for the actual load it carried, the repetition of the same elements leads to the characteristic mesh design of these lattice shells and towers. Putting the aesthetic advantages to one side, this principle considerably reduces the numbers of different parts and connections.

All this shows that, in addition to his virtuosity in a wide range of engineering disciplines, Shukhov also possessed a great talent for rationalisation of the structures he conceived. This integrated approach to solving a technical task is the most prominent feature of his way of working. Shukhov's design method is defined by two central guiding principles. The structures he conceived are always remarkable for using a minimum of material – as dictated by necessity in Russia at the turn of the century. Moreover, economic purposefulness is at the heart of these structures, which manifests itself in rational design processes, a high degree of prefabrication and short assembly times. The consistently pragmatic implementation

Hyperbolic structures: Shukhov's lattice towers – forerunners of modern lightweight construction, First Edition. Matthias Beckh.
© 2015 John Wiley & Sons, Ltd. Published 2015 by John Wiley & Sons, Ltd.

1

of these guiding principles results in radical, novel forms of construction with no directly related predecessors in building history. These guiding principles combine with his extraordinary sense of aesthetics to result in Shukhov's uniqueness. His synthesis of engineering, economy and aesthetics is nowhere so apparent in Shukhov's work than in the NiGRES tower on the Oka, the construction and structural behaviour of which is examined in the chapter of the same name (see p. 94ff.). The effects on architecture of Shukhov's hyperbolic lattice towers should also not be underestimated. Together with Shukhov's other contributions to weight-optimised construction, such as his tied arches, suspended roofs and lattice shells, his works helped initiate modern lightweight construction and influenced developments right up to the present day. How very advanced Shukhov was of his time is illustrated by the fact that many of the designs he developed have been taken up again – more than 100 years later – well after they had seemed to be forgotten.

In this respect, a renaissance for hyperbolic lattice towers and structures would also be highly appreciated. By varying the basic parameters, designers can create an almost unlimited variety of forms. In combination with favourable load-carrying behaviour, the possible applications are many and varied, extending from lookout towers and various types of masts to high-rise building structures.

Remaining questions and outlook

The analysis of the development of Shukhov's water towers is based on the material in the named archives available to view at the time of writing this book. It can be expected that the consideration and evaluation of further sources as part of a current research project will yield more discoveries about the design process and the development of the towers. This applies in particular to the period between 1916 and 1930 for which few examples exist. Furthermore, an analysis of the constructional details and their development remains to be carried out, a task which will require extensive structural survey work on site. In this context, material testing to identify the steels used is currently be performed to give more conclusive proof of their mechanical and metallurgical properties.

In the field of engineering research, the question of the wind loads to be applied in the analysis of these structures is of particular importance. As already mentioned, the wind loads in the calculations investigated in this book were in accordance with DIN standards. Because of the complex three-dimensional geometry of hyperbolic lattice structures, it is impossible to use the shielding factors in international wind standards, because these relate only to lattice structures with orthogonal or triangular plan shapes. Therefore no reduction factor for shielding is applied to the member loads from wind. As a result, the member wind loads calculated from the structural analysis of the NiGRES tower are too large and do not reflect reality. Therefore, currently undertaken wind tunnel tests explore the flow behaviour and shielding effects of the lattice. Only with the help of the precise values from the material properties tests and the wind tunnel tests will it be possible exactly to determine the present safety factors of Shukhov towers.

Further research is required in the use of this form of construction in real projects. One example is an analysis of the extent to which hyperbolic lattice structures are more efficient in high-rise buildings than trussed tubes or diagrid structures. Another field of use could be in mast structures for wind turbines. As nacelle heights increase, the traditional tube cross sections reach their limits for transport, while conventional braced frame towers are often met with objections in populated areas. Therefore hyperbolic lattice structures, with their sophisticated and visually appealing shape could find application in this field.

Towers in comparison

In the archive of the Russian Academy of Sciences in Moscow can be found two sets of tables prepared by the construction company Bari that list Shukhov's water towers and their most important data. The first table (RAN-1508-79/1) contains the data from 28 water towers from the period between 1896 and 1914 (p. 136ff.). In the second table (RAN-1508-79/2) are summaries of the data from ten further towers (p. 140f.). The tables are not dated, but comparison with other sources (e.g. dated blueprints) leads to the conclusion that this collection of towers was built along the Central Asian Railway in about 1915 [1].

In addition to the geometric information about the load-bearing structure, such as height, radii as well as the type and number of vertical members in the lattice, the tables are most notable for the details they give about the water tanks. Their construction type (flat bottomed, suspended bottom or Intze tanks), capacity, equipment and further special aspects such as insulation and the like are precisely listed. The tables also contain further information about the weights of the steel components, plates and rivets as well as details of the materials and method of construction of the foundations. Using the data on the geometry, tank characteristics and weights in the tables, it was possible for the designer to quickly estimate the amount of steel required for and therefore the costs of new water tower projects.

In order to analyse the development of Shukhov's water towers, the tables were translated and the units of measurement converted to the metric system. The parameters defining the form of the structures, which were height, number of vertical members and rotation angle, were examined in relation to the tank size as the basis for the investigations. The values are given for a number of water towers in two charts (p. 120f.). The evaluation of the material and the interactions of the various factors can be found on p. 90ff. Because the tables give no indication of the rotation angle φ, it had to be deduced from images (drawings or photographs): with a knowledge of the number of vertical members, it is possible to work out the central angle; the rotation angle can be calculated from this value and the number of intersections of the straight members. The number and spacing of the horizontal intermediate rings also had to be measured or – in the case of photographs – estimated. Even so, drawings or photographs were not available for all of the towers in the tables, which meant that a total of only 16 could be evaluated. To this are added two further towers in Tyumen and Dnipropetrovsk, whose details were taken from Dmitrij Petrov's book. [2] These 18 towers are redrawn in elevation, plan and as an axonometric projection (scale 1:333 1/3) (p. 122ff.).

The relatively high number of examples for the period between the first completed tower in 1896 in Nizhny Novgorod and 1915 is quite remarkable. At the time of printing however, it was not possible to find tables for later years. Hence only one example was found for the years after 1915, which is the water tower built in Dnipropetrovsk in 1930.

Hyperbolic structures: Shukhov's lattice towers – forerunners of modern lightweight construction, First Edition. Matthias Beckh.
© 2015 John Wiley & Sons, Ltd. Published 2015 by John Wiley & Sons, Ltd.

Water tower in Nizhny Novgorod (RUS) 1896

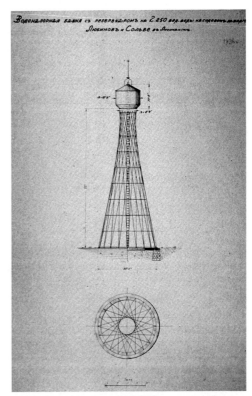

Water tower in Lysychansk (UA) 1896

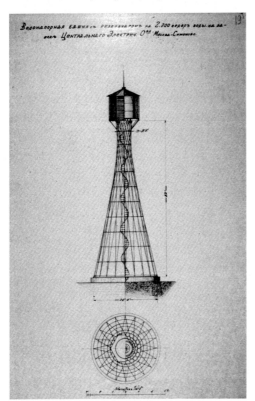

Water tower in Moscow Simonov (RUS) 1899

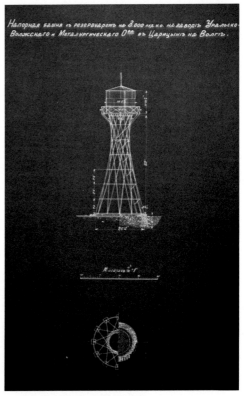

Water tower in Tsaritsyn (RUS) 1899

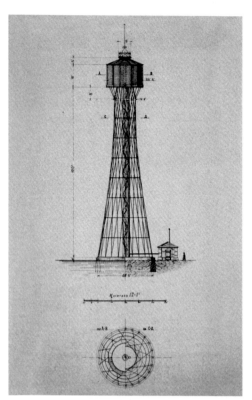

Water tower in Kolomna (RUS) 1902

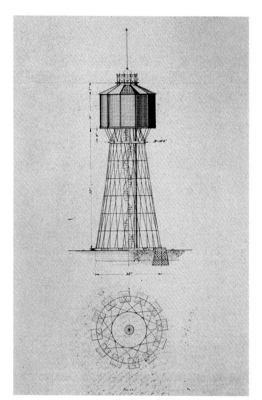

Water tower in Yefremov (RUS) 1902

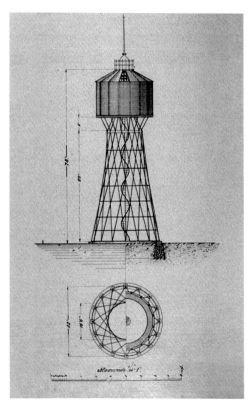

Water tower in Yaroslavl (RUS) 1904

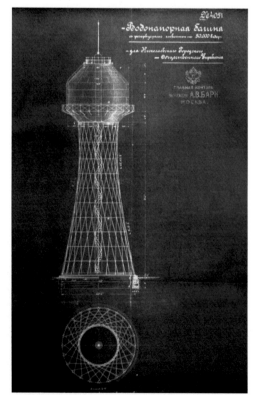

Water tower in Mykolaiv (UA) 1907

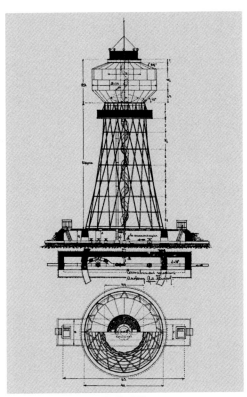

Water tower in Tyumen (RUS) 1908

Water tower in Andijan (UZ) 1909

Water tower in Sagiri (AZ) 1912

Water tower in Kharkiv (UA) 1912

Water tower in Samarkand (UZ) 1913

Water tower in Pryluky (UA) 1914

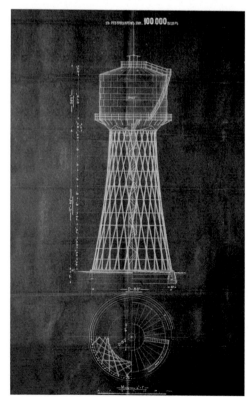

Water tower in Voronezh (RUS) 1915

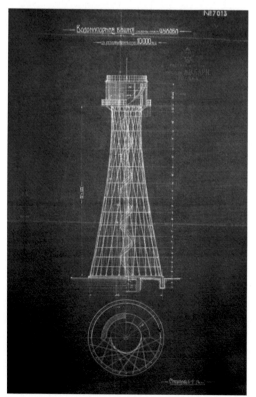

Water tower in Kazalinsk (RUS) 1915

Water tower in Tambov (RUS) 1915

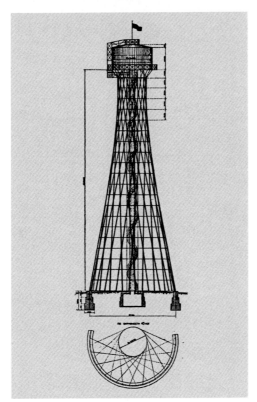

Water tower in Dnipropetrovsk (UA) 1930

Summary of tower height, rotation angle, number of vertical members and tank volume

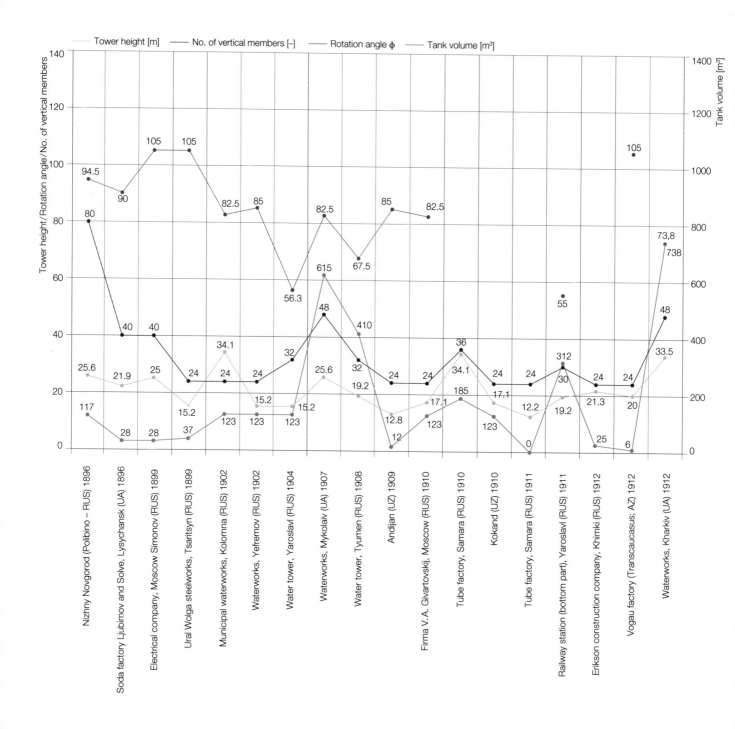

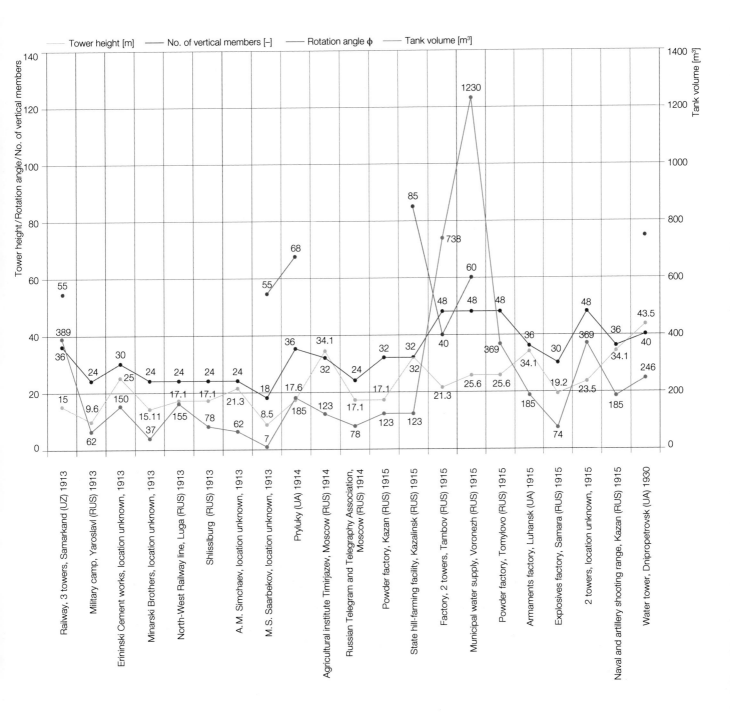

Tower height [m] — No. of vertical members [–] — Rotation angle φ — Tank volume [m³]

Nizhny Novgorod (RUS) 1896

First hyperbolic lattice tower for the All-Russia Exhibition

Lysychansk (UA) 1896

Water tower for the Ljubimov and Solve soda factory

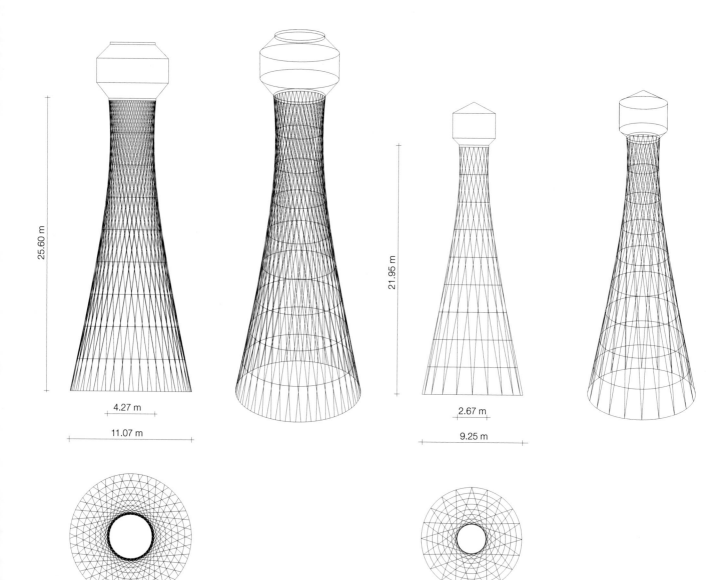

25.60 m

4.27 m

11.07 m

21.95 m

2.67 m

9.25 m

V = 117 m³   φ = 94.5°
2n = 80 (L76/76/10 mm)

V = 27.7 m³   φ = 90°
2n = 40 (L76/76/6 mm)

Towers in comparison

Moscow Simonov (RUS) 1899

Water tower for an electrical company

Tsaritsyn (RUS) 1899

Water tower of the Ural Wolga steelworks

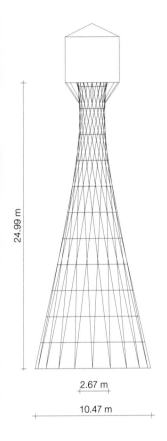
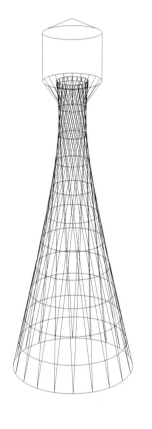

24.99 m

2.67 m

10.47 m

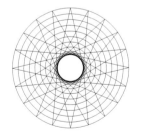

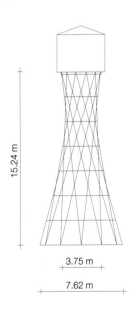
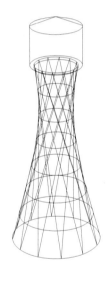

15.24 m

3.75 m

7.62 m

V = 28.3 m³   φ = 105°
2n = 24 (L76/76/6 mm)

V = 36.9 m³   φ = 105°
2n = 24 (L76/76/6 mm)

Kolomna (RUS) 1902

Water tower of the municipal waterworks

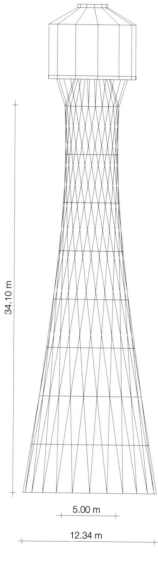
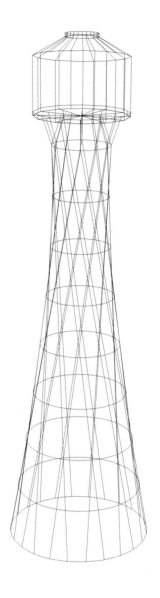

34.10 m

5.00 m

12.34 m

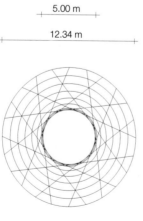

$V = 123\ m^3 \qquad \varphi = 82.5°$
$2n = 24\ (L\,150/150/15\ mm)$

Yefremov (RUS) 1902

Water tower for a waterworks

Yaroslavl (RUS) 1904

Water tower

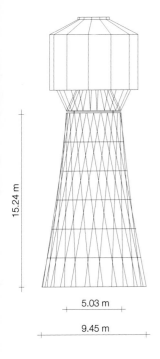

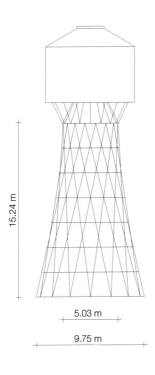

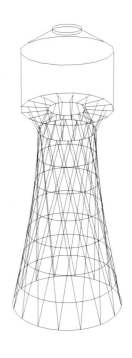

15.24 m

15.24 m

5.03 m

9.45 m

5.03 m

9.75 m

$V = 123\ m^3 \qquad \varphi = 85°$
$2\,n = 24$ (L120/120/12 mm)

$V = 123\ m^3 \qquad \varphi = 56.3°$
$2\,n = 32$ (L100/100/10 mm)

Mykolaiv (UA) 1907

Water tower for a waterworks

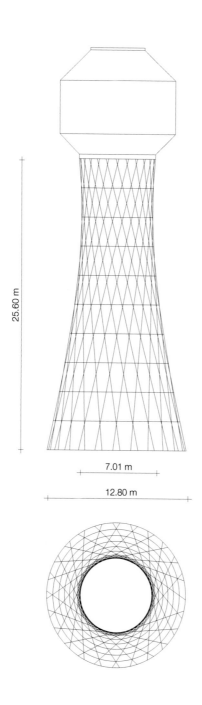

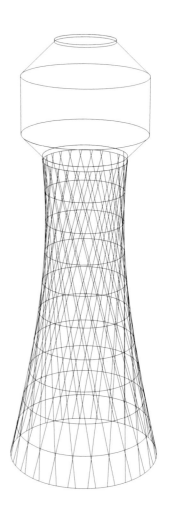

25.60 m

7.01 m

12.80 m

$V = 615 \, m^3$ $\varphi = 82.8°$
$2n = 48$ (L 120/120/12 mm)

Tyumen (RUS) 1908

Water tower

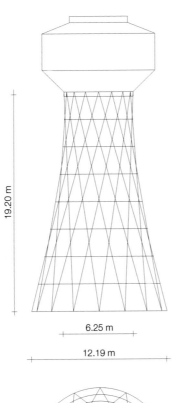

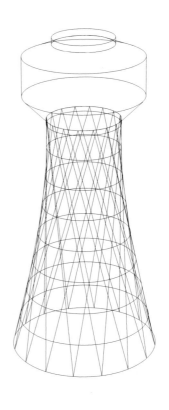

19.20 m

6.25 m

12.19 m

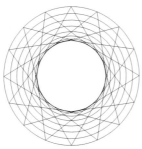

$V = 410 \, m^3$ $\varphi = 67.5°$
$2n = 32$ (∟110/110/12 mm)

Andijan (UZ) 1909

Water tower

Sagiri (AZ) 1912

Water tower for the Vogau factory

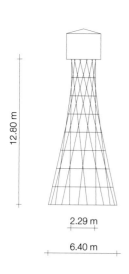

12.80 m

2.29 m

6.40 m

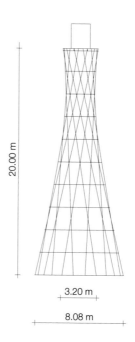
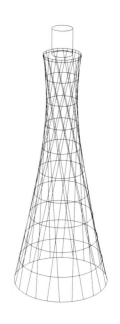

20.00 m

3.20 m

8.08 m

V = 12.3 m³   φ = 85°
2n = 24 (L76/76/6 mm)

V = 6 m³   φ = 105°
2n = 24 (L60/60/6 mm)

Towers in comparison

Kharkiv (UA) 1912

Water tower for a waterworks

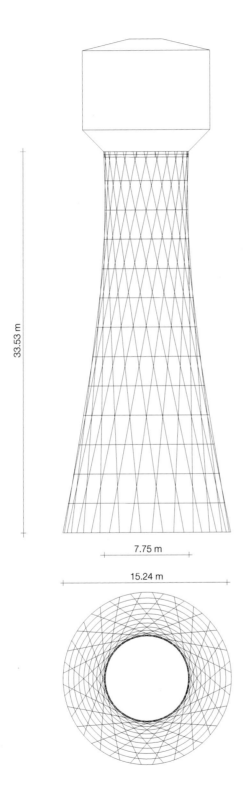

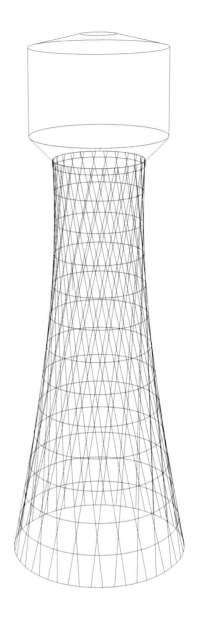

33.53 m

7.75 m

15.24 m

V = 738 m³   φ = 73.8°
2 n = 48 (2× U-section 140 mm)

Samarkand (UZ) 1913

Water tower alongside a railway line

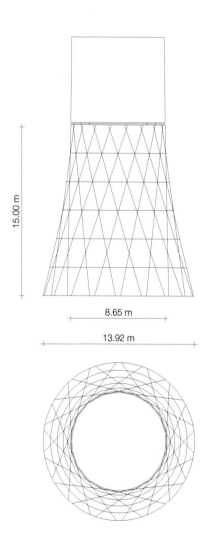
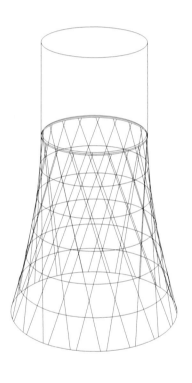

15.00 m

8.65 m

13.92 m

V = 388.5 m³   φ = 55°
2n = 36 (2× U-section 100 mm)

Pryluky (UA) 1914

Water tower

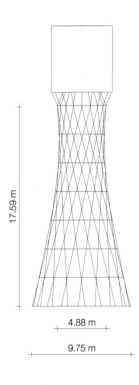

17.59 m

4.88 m

9.75 m

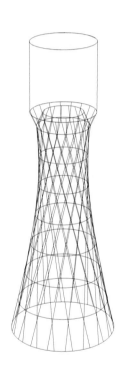

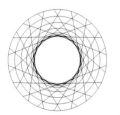

$V = 184.5 \ \text{m}^3$ $\varphi = 68°$
$2n = 24$ (∟100/100/10 mm)

Voronezh (RUS) 1915

Water tower for a municipal water supply

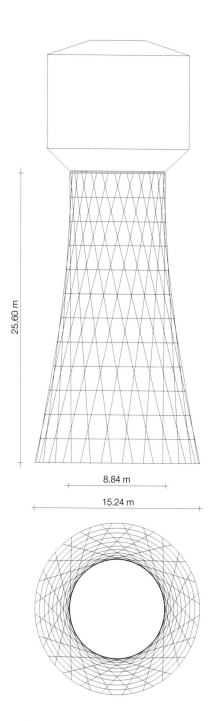
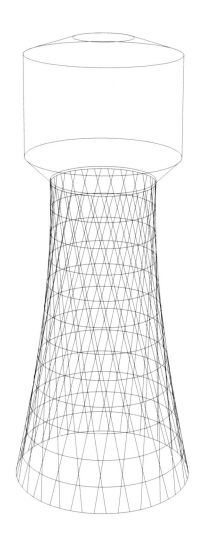

25.60 m

8.84 m

15.24 m

$V = 1230 \text{ m}^3$ $\varphi = 60°$
$2\,n = 48$ (2× U-section 160 mm)

Kazalinsk (RUS) 1915

Water tower for a state hill-farming facility

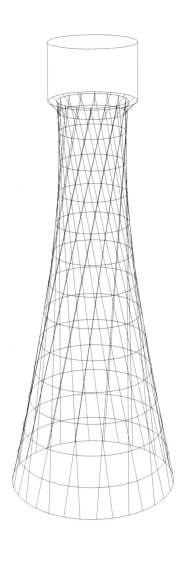

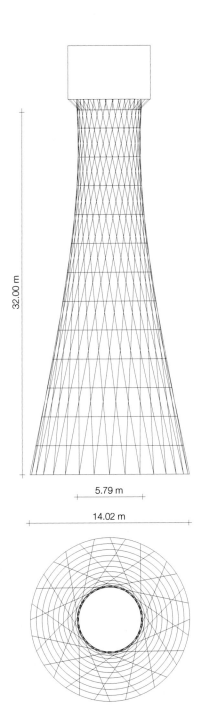

32.00 m

5.79 m

14.02 m

V = 123 m³   φ = 85°
2n = 32 (L100/100/12 mm)

Tambov (RUS) 1915

Water tower for a factory

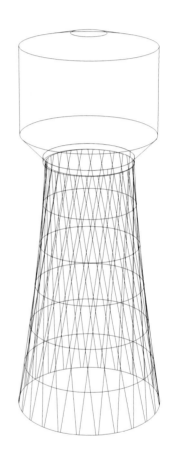

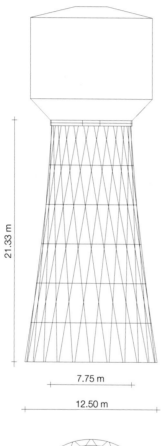

21.33 m

7.75 m

12.50 m

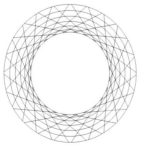

$V = 738 \ m^3$ $\varphi = 40°$

$2n = 48$ (2× U-section 140 mm)

Dnipropetrovsk (UA) 1930

Water tower

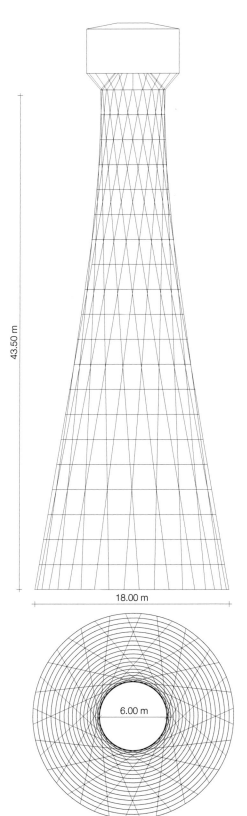

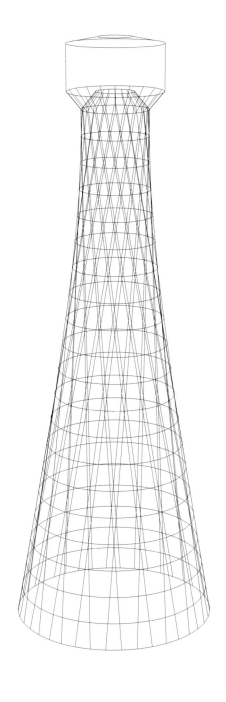

43.50 m

18.00 m

6.00 m

V = 246 m³   φ = 75°
2n = 40 (L100/100/12 mm)

Translation and analysis of the original Russian table RAN OP 1508-79/1 from the archive of the Russian Academy of Sciences in Moscow

		Nizhny Novgorod (Polibino)	Lysychansk, soda factory Ljubimov and Solve	Moscow Simonov, electrical company	Tsaritsyn, Ural Wolga steelworks	Kolomna, municipal waterworks	Yefremov, waterworks	Yaroslavl
		1896	1896	1899	1899	1902	1902	1904
Height of load-bearing structure [m]		25.60	21.95	24.99	15.24	36.58/34.14	17.07/15.24	17.07/15.24
Reservoir capacity [m³]		116.85	27.68	28.29	36.90	123.00	123.00	123.00
Reservoir								
Diameter [m]	Tank	6.55	3.89	3.17	4.42	6.25	6.25	6.25
	Internal pipe	1.42	0.76			1.22	1.22	1.22
Height [m]	Conical roof	1.22	0.86					
	Cylindrical part	2.14/5.03	2.00	3.61	3.63	4.24	4.24	4.24
	Bottom part	1.37	0.61					
Construction		Intze principle No cladding	Intze principle No cladding	Flat bottom Timber cladding	Flat bottom No cladding	Flat bottom Timber cladding	Flat bottom Timber cladding	Flat bottom Timber cladding
Load-bearing structure								
Diameter bottom [m]		11.07	9.25	10.47	7.62	12.34	9.45	9.75
Diameter top [m]		4.267	2.67	2.67	3.75	5.00	5.03	5.03
No. of vertical lattice members		80	40	40	24	24	24	32
Cross section [mm]		∟76.2/76.2/9.5 ∟50.8/50.8/7.9	∟76.2/76.2/6.4 ∟50.8/50.8/6.4	∟76.2/76.2/6.4 ∟50.8/50.8/6.4	∟6.2/76.2/6.4	∟150/150/15 ∟120/120/15 ∟120/120/14 ∟120/120/12	∟120/120/12 ∟120/120/10	∟101.6/101.6/9.5
No. of bolts × diameter [mm]		20× 31.8	20× 31.8	20× 38.1	12× 25.4	24× 38.1	12× 31.8	16× 34.9
Foundation								
Sheets/plates in iron [mm]		–	–	–	–	Cast-iron plates 24× 457.2/457.2/ 31.8	Cast-iron plates 12× 609.6/457.2/ 44.5	Cast-iron plates 16× 558.8/457.2/ 31.8
Height [m]		2.13	1.83	2.85	1.83	3.05	2.44 (twelve separate foundations)	2.44
Length top [m]		1.02	0.71	0.76	0.71	0.71	1.42	0.71
Length bottom [m]		1.22	1.42	1.52	1.42	1.42	2.13	1.42
Weight								
Reservoir [kN]		83.37	34.40	19.66	29.16	58.80	60.61	57.33
Structure [kN]		296.64	87.63	129.40	55.53	421.95	170.02	186.90
Total weight [kN]		380.02	122.03	149.06	84.69	480.75	230.63	244.23

Mykolaiv, waterworks	Andijan	Moscow, V. A. Givartovsky	Samara, tube factory	Kokand	Samara, tube factory	Yaroslavl, railway station	
1907	1909	1910	1910	1910	1911	1911	
25.60	12.80	17.07	36.58/34.14	17.07	12.19	20.27 + 19.20 (upper + lower)	**Height of load-bearing structure [m]**
615.00	12.30	123.00	184.50	123.00	0.37	194.25 + 116.55 (upper + lower)	**Reservoir capacity [m³]**
							Reservoir
10.52 1.07	2.59	5.49	6.45 1.20	5.49	0.76	7.47 + 5.87 1.22 + 1.22 (upper + lower)	Diameter [m]
2.74 4.80 1.75	2.44	4.78 0.91	5.99	4.78 0.91	0.91	(upper + lower) 4.13 + 4.20 1.25 + 0.91	Height [m]
Intze principle No cladding	Flat bottom No cladding	Spherical bottom Timber cladding	Flat bottom Timber cladding	Spherical bottom No cladding	Flat bottom No cladding	Spherical bottom With cladding	Construction
							Load-bearing structure
12.80	6.40	9.14	12.34	9.14	4.27	15.70	Diameter bottom [m]
7.01	2.29	5.49	5.03	5.49	2.13	8.53 + 15.70 (upper + lower)	Diameter top [m]
48	24	24	36	24	24	30	No. of vertical lattice members
L127/127/12.7 L101.6/101.6/9.5	L76.2/76.2/6.4 L63.5/63.5/6.4	L101.6/101.6/9.5	L127/127/12,7 L101.6/101.6/9.5	L88.9/88.9/9.5	L50.8/50.8/6.4	L114.3/114.3/15.9/12.7 L101.6/101.6/12.7/11.1 L88.9/88.9/9.5	Cross section [mm]
24× 28.6	12× 25.4	24× 25.4	36× 38.1	24× 25.4	12× 19.1	30× 31.8	No. of bolts × diameter [mm]
							Foundation
304.8/12.7	Cast-iron plates 12× 304.8/304.8/25.4	Cast-iron plates 24× 508/228.6/31.8	–	Cast-iron plates 24× 508/228.6/1.8		–	Sheets/plates in iron [mm]
2.13	2.13 (twelve separate foundations)	2.13	2.44	2.13	Supported on Å beams No. 32, 28, 15	2.13	Height [m]
0.71	0.71	0.71	0.71	0.71		0.71	Length top [m]
1.07	1.07	1.07	1.91	1.07		1.067	Length bottom [m]
							Weight
191.65	12.29	49.14	76.50	76.17	1.80	87.63 + 70.43	Reservoir [kN]
562.16	45.21	154.46	419.00	129.89	39.97	498.28	Structure [kN]
753.81	57.50	203.60	495.50	206.06	41.77	656.35	Total weight [kN]

Translation and analysis of the original Russian table RAN OP 1508-79/1 from the archive of the Russian Academy of Sciences in Moscow

		Sagiri, (Transcaucasus) Vogau factory	Khimki, Erikson construction company	Kharkiv, waterworks	Samarkand, railway 3 towers	Yaroslavl, military camp	Unknown location, Erininski cement works	Unknown location, Mlnarski Brothers
		1912	1912	1912	1913	1913	1913	1913
Height of load-bearing structure [m]		20.00	21.34	33.53	15.00	9.60	25.00	15.11
Reservoir capacity [m³]		6.00	24.60	738.00	388.50	61.50	150.00	36.90
Reservoir								
Diameter [m]	Tank / Internal pipe	1.80	3.84	11.00 / 1.42	8.61	4.70	6.25	4.08
Height [m]	Conical roof / Cylindrical part / Bottom part	2.44	2.13	1.22 / 7.16 / 1.63	5.96 / 1.57	3.62	4.80	2.83
Construction		Flat bottom No cladding	Flat bottom Timber cladding	Intze principle No cladding	Spherical bottom No cladding	Flat bottom No cladding	Flat bottom Timber cladding	Flat bottom Timber cladding
Load-bearing structure								
Diameter bottom [m]		8.08	9.53	15.24	13.92	6.25	11.72	7.32
Diameter top [m]		3.20	3.51	7.75	8.65	3.66	5.18	3.66
No. of vertical lattice members		24	24	48	36	24	30	24
Cross section [mm]		L 63.5/63.5/6.4	L 63.5/63/7.9 L 63.5/63/6.4	2× fi No. 14	2× fi No. 10	L 63.5/63.5/6.4	L 101.6/101.6/9.5	L 63.5/63.5/7.9
No. of bolts × diameter [mm]		24× 25.4	24× 25.4	48× 38.1	18× 31.8	12× 25.4	30× 25.4	12× 25.4
Foundation								
Sheets/plates in iron [mm]		Cast-iron plates 24× 228.6/228.6/ 25.4	Cast-iron plates 24× 228.6/228.6/ 25.4	317.5/15.9	Cast-iron plates 18× 533/533/9.5	Cast-iron plates 12× 304.8/254/9.5	Cast-iron plates 30× 457.2/304.8/12.7	Cast-iron plates 12× 304.8/304.8/9.5
Height [m]		1.60	1.83	3.20	2.13	1.83	2.13	1.83
Length top [m]		0.53	0.71	6.50	0.71	0.56	0.56	0.71
Length bottom [m]		0.71	0.86	1.22	1.25	0.71	0.89	0.91
Weight								
Reservoir [kN]		9.34	21.46	339.89	151.68	31.45	46.03	19.00
Structure [kN]		66.50	73.38	911.06	222.11	37.18	247.17	61.92
Total weight [kN]		75.84	94.84	1250.94	373.79	68.63	293.20	80.92

Luga, North-West Railway line	Shlisslburg	Location unknown, A. M. Simchaev	Location unknown, M. S. Saarbekov	Moscow, Timirjazev Agricultural Institute	Pryluky	Moscow, Russian Telegram and Telegraphy Association	
1913	1913	1913	1913	1914	1914	1914	
17.07	17.07	21.34	8.53	34.14	17.59	17.07	**Height of load-bearing structure [m]**
77.70 + 77.70 = 155.40	77.70	61.50	7.38	123.00	184.50	77.70	**Reservoir capacity [m³]**
							Reservoir
6.51	5.04	4.71	2.31	6.13	6.43	5.04	Diameter [m]
4.81	3.60 0.91	3.62	1.83	4.24	5.99	3.59 0.91	Height [m]
Flat bottom Timber cladding	Spherical bottom Timber cladding	Flat bottom No cladding	Flat bottom No cladding	Flat bottom Timber cladding	Flat bottom Timber cladding	Spherical bottom Timber cladding	Construction
							Load-bearing structure Diameter bottom [m]
11.58	8.64	9.80	4.66	14.02	9.75	8.64	
6.50	5.11	4.06	2.31	5.79	4.88	5.06	Diameter top [m]
24	24	24	18	32	36	24	No. of vertical lattice members
L101.6/101.6/9.5	L101.6/101.6/9.5	L76.2/76.2/7.9	L63.5/63.5/6.4	2× fi No. 8	L101.6/101.6/9.5	L101.6/101.6/9.5	Cross section [mm]
24× 25.4	24× 25.4	24× 25.4	18× 15.9	32× 25.4	36× 25.4	24× 25.4	No. of bolts × diameter [mm]
Cast-iron plates 24× 609.6/406.4/38.1	Cast-iron plates 24× 508/228.6/38.1	Cast-iron plates 24× 304/304/25.4	Cast-iron plates 18× 203.2/203.2/ 12.7	Cast-iron plates 32× 406.4/330.2/ 12.7	–	Cast-iron plates 24× 508/228.6/31.8	**Foundation** Sheets/plates in iron [mm]
2.44	2.13	1.83	1.52	1.83	2.13	2.13	Height [m]
0.71	0.56	0.71	0.41	0.56	0.56	0.56	Length top [m]
1.07	0.89	0.86	2.00	0.71	0.71	0.89	Length bottom [m]
							Weight
83.21	36.20	31.45	8.35	40.30	66.83	36.20	Reservoir [kN]
190.01	135.46	127.60	27.36	390.17	221.29	144.47	Structure [kN]
273.22	171.66	159.05	35.71	430.47	288.12	180.67	Total weight [kN]

Translation and analysis of the original Russian table RAN OP 1508-79/2 from the archive of the Russian Academy of Sciences in Moscow

	Kazan, powder factory	Kazalinsk, state hill-farming facility	Unknown location, public limited company P. W. Baranov	Tambov factory 2 towers	Voronezh, municipal water supply
	1915	1915	1915	1915	1915
	Drawing No. 3621	Drawing No. 7013	Drawing No. 113	Drawing No. 7077	Drawing No. 7084
Height of load-bearing structure [m]	17.07	32.00	33.53	21.34	25.60
Reservoir capacity [m³]	123.00	123.00	738.00	738.00	1230.00
Reservoir					
Diameter [m]	6.25	6.10	11.00	11.00	12.80
Height [m]	4.24	4.24	7.01	7.010	8.36
Construction	Flat bottom, Internal pipe	Flat bottom	Intze principle	Intze principle	Intze principle
Grillage under the reservoir	16× Å No. 18	16× Å No. 15			
Load-bearing structure					
Diameter bottom [m]	9.75	14.02	15.24	12.50	15.24
Diameter top [m]	4.12	5.79	7.75	7.75	8.84
No. of vertical lattice members	32	32	48	48	48
Cross section [mm]	L101.6/101.6/9.5	L101.6/101.6/12.7 L101.6/101.6/9.5	2×fi No. 14	2×fi No. 14	2×fi No. 16
No. of bolts × diameter [mm]	16× 31.8	32× 25.4	48× 25.4	48× 25.4	48× 25.4
Foundation					
Length top [m]	0.71	0.56	0.71	0.71	0.71
Length bottom [m]	1.42	0.71	1.22	1.22	1.52
Height [m]	2.44	2.13	2.44	2.44	2.44
Volume of masonry [m³]	79.57	59.47	112.70	2× 112.70 = 225.40	130.26
Weight					
Anchor bolts [kN]	2.00 + 12.07	5.85	8.65	2× 8.55 = 17.10	5.86
Support ring [kN]	4.96	14.91	48.51	2× 41.39 = 82.78	47.66
Vertical members [kN]	85.91	216.61	584.96	2× 395.49 = 790.98	546.86
Horizontal rings [kN]	15.56	44.39	73.08	2× 49.55 = 99.10	85.84
Top ring [kN]	6.18	5.57	–	–	–
Reservoir [kN]	56.86	40.20	–	–	–
Intze tank with roof and top ring [kN]	–	–	317.01	2× 317 = 634.00	507.32
Spiral staircase from ground level [kN]	6.98	33.82	21.83	15.12	32.55
Maintenance platform below the reservoir [kN]	–	–	47.75	2× 34.71 = 69.42	49.39
Crossing platform below the reservoir, bridge and steps [kN]	–	4.72	–	–	–
Grillage for reservoir [kN]	43.72	21.89	–	–	–
Reservoir circumference [kN]	–	9.31	38.69	2× 26.75 = 53.50	35.57
Steps, railings and ladder [kN]	6.05	15.85	22.89	49.99	34.60
Total weight [kN]	240.29	413.11	1163.36	1818.94	1345.65
Rivets in the structure, excluding reservoir [kN]	4.45	11.73	20.05	26.67	26.60

Tomylovo, powder factory	Luhansk, armaments factory	Samara, explosives factory	Unknown location, 2 towers	Naval and artillery shooting range	
1915	1915	1915	1915	1915	
Drawing No. 7059	Drawing No. 7181	Drawing No. 7241	Drawing No. 7201	Drawing No. 73	
25.60	34.14	19.20	23.47	34.14	**Height of load-bearing structure [m]**
369.00	184.50	73.80	369.0	184.50	**Reservoir capacity [m³]**
					Reservoir
8.53	6.43	4.75	8.13	6.43	Diameter [m]
5.74	5.99	4.19	7.16	5.79	Height [m]
Intze principle	Flat bottom, Internal pipe	Flat bottom	Flat bottom	Flat bottom	Construction
	18× Å No. 20	15× Å No. 15	24× Å No. 25	18× Å No. 2	Grillage under the reservoir
					Load-bearing structure
12.20	12.34	9.16	12.19	12.34	Diameter bottom [m]
6.10	5.03	4.58	6.10	5.029	Diameter top [m]
48	36	30	48	36	No. of vertical lattice members
∟101.6/101.6/12.7 ∟101.6/101.6/9.5	∟101.6/101.6/15.9 ∟101.6/101.6/12.7 ∟101.6/101.6/9.5	∟88.9/88.9/9.5	∟101.6/101.6/12.7	∟101.6/101.6/15.9 ∟101.6/101.6/12.7 ∟101.6/101.6/9.5	Cross section [mm]
48× 25.4	36× 25.4	30× 25.4	48× 25.4	36× 25.4	No. of bolts × diameter [mm]
					Foundation
0.56	0.56	0.56	0.56	0.56	Length top [m]
1.07	0.91	0.89	1.07	0.91	Length bottom [m]
2.44	2.44	2.44	3.35	2.44	Height [m]
68.67	69.66	50.69	94.58	69.66	Volume of masonry [m³]
					Weight
8.65	5.22	3.10	7.18	5.22	Anchor bolts [kN]
15.81	15.14	8.32	18.55	15.14	Support ring [kN]
241.83	266.42	86.20	253.02	266.42	Vertical members [kN]
44.80	33.58	22.87	44.93	33.58	Horizontal rings [kN]
–	6.13	5.03	8.13	5.80	Top ring [kN]
–	76.46	31.00	89.07	66.77	Reservoir [kN]
171.31	–	–	–	–	Intze tank with roof and top ring [kN]
17.66	31.43	11.08	33.39	47.35	Spiral staircase from ground level [kN]
24.37	–	–	–	–	Maintenance platform below the reservoir [kN]
–	–	7.91	14.19	14.81	Crossing platform below the reservoir, bridge and steps [kN]
–	57.59	28.15	75.55	46.66	Grillage for reservoir [kN]
22.57	–	–	26.79	–	Reservoir circumference [kN]
17.42	4.70	9.11	12.50	2.19	Steps, railings and ladder [kN]
564.43	496.70	212.76	583.31	503.86	Total weight [kN]
12.92	12.50	5.43	11.99	11.93	Rivets in the structure, excluding reservoir [kN]

Notes

Introduction

[1] Konfederatov, I. Ja.: Vladimir Grigor'evič Šuchov. Moscow/Leningrad 1950
[2] Lopatto, A. E.: Vladimir Grigor'evich Šuchov – vydajuščijsja russkij inžener. Moscow 1951
[3] Išlinskij, A. Ju.: Über Šuchovs Beitrag zur Planung und Berechnung von Baukonstruktionen (Transliteration into German by Ottmar Pertschi). In: Šuchov, Vladimir G.: Stroitel'naja mechanika. Izbrannye trudy. Moscow 1977. pp. 4–9. Translation Service of the University Library Stuttgart. Ü/247.
[4] Graefe, Rainer; Gappoev, Murat; Pertschi, Ottmar: Vladimir G. Šuchov. 1853–1939. Die Kunst der sparsamen Konstruktion. Stuttgart 1990
[5] Tomlow, Jos: Zur Formfindung bei Šuchovs mehrstöckigen Gittertürmen aus Hyperboloiden. In: Vom Holz zum Eisen. Weitgespannte Konstruktionen des 18. und 19. Jahrhunderts. German-Soviet Colloquium of the Mittel-, Südost-, Osteuropa-Referats des Instituts für Auslandsbeziehungen, Stuttgart and Sub-project C3 "Geschichte des Konstruierens" in the Collaborative Research Centre 230 "Natürliche Konstruktionen" on 25th and 26th January 1990 at Institute for Lightweight Structures and Conceptual Design at the University of Stuttgart, 1991, pp. 166–185
[6] Günther, Daniel: Die hyperbolischen Gitterstabkonstruktionen von V. G. Šuchov. Degree thesis at the Technische Universität München, 2003
[7] De Vries, Peter: Morphology and structural behavior of the hyperbolic lattice. In: 4. International Colloquium on Structural Morphology. Delft University of Technology, 2000

Building with hyperbolic lattice structures

[1] Peters, Tom Frank: Building the Nineteenth Century. Cambridge 1996, p. 42
[2] Peters, Tom Frank: Time is Money – die Entwicklung des modernen Bauwesens. Stuttgart 1981, p. 223
[3] Pfammatter, Urich: In die Zukunft gebaut – Bautechnik- und Kulturgeschichte von der Industriellen Revolution bis heute. Munich 2005, p. 66
[4] Lorenz, Werner: Konstruktion als Kunstwerk – Bauen mit Eisen in Berlin und Potsdam 1797–1850. Berlin 1995, p. 8
[5] Straub, Hans: Die Geschichte der Bauingenieurkunst. Basel 1992, p. 206
[6] ibid. p. 260
[7] Schädlich, Christian: Der Baustoff Eisen als Grundlage für die Herausbildung qualitativ neuer Baukonstruktionen im 19. Jahrhundert. In: Graefe, Rainer (Ed.): Zur Geschichte des Konstruierens. Stuttgart 1989, p. 146
[8] Lorenz, Werner: Die Entwicklung des Dreigelenksystems im 19. Jahrhundert. In: Stahlbau 01/1990, pp. 1–10.
[9] Giedion, Sigfried: Raum Zeit Architektur – die Entstehung einer neuen Tradition. Basel 1996, p. 159
[10] Graefe, Rainer; Gappoev, Murat; Pertschi, Ottmar: Vladimir G. Šuchov. 1853–1939. Die Kunst der sparsamen Konstruktion. Stuttgart 1990
[11] ibid. p. 8
[12] ibid. p. 8
Ricken, Herbert: Erinnerungen an Wladimir Grigorjewitsch Schuchow (1853–1939). In: Bautechnik 08/2003.
[13] Graefe, Rainer: Netzdächer, Hängedächer und Gitterschalen. In: Graefe, Rainer; Gappoev, Murat; Pertschi, Ottmar: Vladimir G. Šuchov. 1853–1939. Die Kunst der sparsamen Konstruktion. Stuttgart 1990, p. 39
[14] ibid. p. 43
[15] Beckh, Matthias; Barthel, Rainer: The first doubly curved gridshell structure – Shukhov's building for the plate rolling workshop in Vyksa, In: Kurrer, Karl-Eugen; Lorenz, Werner; Wetzk, Volker: Proceedings of the Third International Congress on Construction History. Berlin 2009, pp. 159–166.
[16] see also:
Barthel, Rainer et al.: Ein Meilenstein im Schalenbau – Šuchovs Halle für das Blechwalzwerk in Vyksa. In: Mayer, Juliane (ed.): Festschrift für Rainer Graefe – Forschen, Lehren und Erhalten. Innsbruck 2009, pp. 105–122
Barthel, Rainer; Kayser, Christian: Šuchov Halle in Vyksa – Dokumentation des Bestandes und der Schäden. Barthel & Maus – Beratende Ingenieure. 2007 (unveröffentlicht).
[17] Beckh, Matthias: Von der Rippe zum Netz – die Entwicklung der Schalentragwerke aus Eisen und Stahl. In: Nerdinger, Winfried (ed.): Wendepunkte im Bauen. Munich 2010, pp. 38–45.
[18] cf. 13, p. 35
[19] Bach, Klaus: Gittermasten russischer und amerikanischer Schlachtschiffe. In: Graefe, Rainer; Gappoev, Murat; Pertschi, Ottmar: Vladimir G. Šuchov. 1853–1939. Die Kunst der sparsamen Konstruktion. Stuttgart 1990, p. 104
[20] cf. 10, p. 177
[21] ibid.
[22] Petropavlovskaja, Irina A.: Hyperbolische Gittertürme. In: Graefe, Rainer; Gappoev, Murat; Pertschi, Ottmar: Vladimir G. Šuchov. 1853–1939. Die Kunst der sparsamen Konstruktion. Stuttgart 1990, p. 78
[23] ibid. p. 78
[24] English, Elizabeth Cooper: Arkhitektura I Mnimosti: the origins of Soviet Avant-Garde rationalist architecture in the Russian mystical-philosophical and mathematical intellectual tradition. University of Pennsylvania. Philadelphia 2000
[25] see also:
Brandt-Mannesmann, Ruthilt: Dokumente aus dem Leben der Erfinder. Remscheid 1964
Burkhardt, Berthold: Eiserne Rohrkonstruktionen im 19. Jahrhundert. In: Vom Holz zum Eisen. Weitgespannte Konstruktionen des 18. und 19. Jahrhunderts. German-Soviet Colloquium of the Mittel-, Südost-, Osteuropa-Referats des Instituts für Auslandsbeziehungen, Stuttgart and Sub-project C3 "Geschichte des Konstruierens" in the Collaborative Research Centre 230 "Natürliche Konstruktionen" on 25th and 26th January 1990 at Institute for Lightweight Structures and Conceptual Design (ILEK) at the University of Stuttgart, 1991, pp. 152–164
[26] Information from Ekaterina Nozhova
[27] cf. 22, p. 78
[28] ibid. p. 78
[29] Daniel Günther: Die hyperbolischen Gitterstabkonstuktionen von V. G. Šuchov. degree thesis at the Technischen Universität München, 2003, p. 97ff.
[30] cf. 10, p. 18

Geometry and form of hyperbolic frameworks

[1] Martins, Luis da Motta Faria Câncio: Morphologie der gekrümmten Flächentragwerke. Dissertation at the ETH Zurich, Institute of Structural Engineering. 1996, p. 40
[2] Meyberg, Kurt; Vachenauer, Peter: Höhere Mathematik 1. Heidelberg 1993, p. 344f.
Bronštejn, Il'ja Nikolaevič et al.: Taschenbuch der Mathematik. Frankfurt/Main 1993, p. 162f.
[3] Meyberg, Kurt; Vachenauer, Peter: Höhere Mathematik 1. Heidelberg 1993, p. 344
[4] Bronštejn, Il'ja Nikolaevič et al.: Taschenbuch der Mathematik. Frankfurt/Main 1993, p. 167
[5] http://mathworld.wolfram.com/Hyperboloid.html. Version 23.07.2012
[6] ibid.
[7] Hoheisel, Meike: Einfluss der Formparameter auf das Tragverhalten hyperbolischer Stabwerke. Master Theses on the subject of civil engineering, Chair of Structural Design/Chair of Structural Analysis, Technische Universität München, 2010, p. 11

Structural analysis and calculation methods

[1] Kollar, Lajos: Die dehnungslosen Formänderungen von Schalen. In: Zerna, Wolfgang: Konstruktiver Ingenieurbau – Berichte. Heft 20. Das Problem der dehnungslosen Verformungen. Essen 1974, pp. 35
[2] Bletzinger, Kai-Uwe: Theory of Shells. Lecture notes summer semester 2005, Chair of Structural Analysis, Technische Universität München. p. I–18.
[3] Sumec, Jozef: General Stability Analysis of Lattice Shells by Continuum Modelling. In: International Journal of Space Structures. Vol. 7, No. 4/1992, pp. 275–283.
[4] Bulenda, Thomas; Knippers, Jan; Sailer, S.: Untersuchungen zum Tragverhalten von Netzkuppeln. In: Ramm, Ekkehard; Stein, Erwin; Wunderlich, Walter: Finite Elemente in der Baupraxis. Modellierung, Berechnung und Konstruktion. Berlin 1995
[5] see also:
Bulenda, Thomas; Knippers, Jan: Stability of gridshells. In: Computers and structures 79/2001, pp. 1161–1174
Knippers, Jan; Bulenda, Thomas; Stein, Michael: Zum Entwurf und zur Berechnung von Stabschalen. In: Stahlbau 01/1997, pp. 31–37
Knippers, Jan: Zum Stabilitätsverhalten tonnenförmiger Stabwerksschalen. In: Stahlbau 04/1998, pp. 298–306
Bulenda, Thomas; Winziger, Thomas: Verfeinerte Berechnung von Gitterschalen. In: Stahlbau 01/2005, pp. 33–38
[6] Graf, Jürgen: Entwurf und Konstruktion von Translationsnetzschalen. Dissertation at the Institute of Structural Design, University of Stuttgart, 2002
[7] Schober, Hans; Kürschner, Kai; Jungjohann, Hauke: Neue Messe Mailand – Netzstruktur und Tragverhalten einer Freiformfläche. In: Stahlbau 08/2004, pp. 541–551.

Hyperbolic structures: Shukhov's lattice towers – forerunners of modern lightweight construction, First Edition. Matthias Beckh.
© 2015 John Wiley & Sons, Ltd. Published 2015 by John Wiley & Sons, Ltd.

[8] see also:
Graf, Jürgen: Entwurf und Konstruktion von Translationsnetzschalen. Dissertation at the Institute of Structural Design, University of Stuttgart, 2002, p. 125f.
Ramm, Ekkehard: Geometrisch nichtlineare Elastostatik und Finite Elemente. Bericht Nr. 76-2. Institute for Structural Mechanics, University of Stuttgart, 1976, p. 79f.

[9] Ramm, Ekkehard: Geometrisch nichtlineare Elastostatik und Finite Elemente. Bericht Nr. 76-2. Institute for Structural Mechanics, University of Stuttgart, 1976, p. 78

[10] Gioncu, Victor; Balut, Nicolae: Instability behavior of single layer reticulated shells. In: International Journal of Space Structures. Vol 7. No 4/1992, pp. 243–252.

[11] Werkle, Horst: Finite Elemente in der Baustatik. Statik und Dynamik der Stab- und Flächentragwerke. Wiesbaden 2008, pp. 37f.

[12] ibid. pp. 41

[13] DIN 18800-2: 2008-11. Stahlbauten. Stabilitätsfälle – Knicken von Stäben und Stabwerken.

[14] see also: Bulenda, Thomas; Knippers, Jan: Stability of gridshells. In: Computers and structures 79/2001; pp. 1161–1174
Knippers, Jan; Bulenda, Thomas; Stein, Michael: Zum Entwurf und zur Berechnung von Stabschalen. In: Stahlbau 01/1997, pp. 31–37.
Bulenda, Thomas; Knippers, Jan; Sailer, S.: Untersuchungen zum Tragverhalten von Netzkuppeln. In: Ramm, Ekkehard; Stein, Erwin; Wunderlich, Walter: Finite Elemente in der Baupraxis. Modellierung, Berechnung und Konstruktion. Berlin 1995
Graf, Jürgen: Entwurf und Konstruktion von Translationsnetzschalen. Dissertation am Institut für Konstruktion und Entwurf. Universität Stuttgart, 2002

[15] Graf, Jürgen: Entwurf und Konstruktion von Translationsnetzschalen. Dissertation at the Institute of Structural Design, University of Stuttgart, 2002, p. 154

[16] cf. 4

[17] cf. 10

[18] see also: Hoheisel, Meike: Einfluss der Formparameter auf das Tragverhalten hyperbolischer Stabwerke. Master Theses on the subject of civil engineering. Chair of Structural Design/Chair of Structural Analysis, Technische Universität München, 2010.
Beckh, Matthias; Hoheisel, Meike: Form und Tragverhalten hyperbolischer Gittertürme. In: Stahlbau 09/2010, pp. 669–681.

Relationships between form and structural behaviour

[1] Hoheisel, Meike.: Einfluss der Formparameter auf das Tragverhalten hyperbolischer Stabwerke. Master Theses on the subject of civil engineering, Chair of Structural Design/Chair of Structural Analysis, Technische Universität München 2010.

[2] Beckh, Matthias; Hoheisel, Meike: Form und Tragverhalten hyperbolischer Gittertürme. In: Stahlbau 09/2010, pp. 669–681.

[3] Gioncu, Victor; Balut, Nicaolae: Instability behavior of single layer reticulated shells. In: International Journal of Space Structures. Vol 7. No 4/1992, pp. 243–252.

[4] Bulenda, Thomas; Winziger, Thomas: Verfeinerte Berechnung von Gitterschalen. In: Stahlbau 01/2005, pp. 33–38

[5] Petrov, Dmitrij V.: Železnye vodonapornye bašni. Ich naznačenie, konstrukcii i rasčety (Transliteration into German: Eiserne Wassertürme. Ihre Bedeutung, Konstruktion und Berechnung). Mykolaiv 1911

[6] Djadjuša, V. A.: Padenie vodonapornoj bašni (Transliteration into German: Der Sturz eines Wasserturms). In: Sanitarnaja technika. Nr. 2–3/1931, pp. 25–30

[7] cf. 1, pp. 99

Design and analysis of Shukhov's towers

[1] Kottenmaier, Eduard: Der Stahlbehälterbau. In: Stahlbau 02/1930, pp. 17–22
Kottenmaier, Eduard: Der Stahlbehälterbau. In: Stahlbau 05/1930, pp. 49–55.

[2] Werth, Jan: Ursachen und technische Voraussetzungen für die Entwicklung der Wasserhochbehälter. In: Becher, Bernd und Hilla: Industriearchitektur des neunzehnten Jahrhunderts. Die Architektur der Förder- und Wassertürme. Munich 1971
Merkl, Gerhard et al.: Historische Wassertürme. Beiträge zur Technikgeschichte von Wasserspeicherung und Wasserversorgung. Munich/Vienna 1985

[3] Werth, Jan: Ursachen und technische Voraussetzungen für die Entwicklung der Wasserhochbehälter. In: Becher, Bernd und Hilla: Industriearchitektur des neunzehnten Jahrhunderts. Die Architektur der Förder- und Wassertürme. Munich 1971, p. 349

[4] ibid. p. 351

[5] ibid. p. 355

[6] ibid. p. 357

[7] ibid. p. 364

[8] ibid. p. 367

[9] ibid. p. 358

[10] ibid. p. 358

[11] see also: Petrov, Dmitrij: Železnye vodonapornye bašni. Ich naznačenie, konstrukcii i rasčety (Transliteration into German: Eiserne Wassertürme. Ihre Bedeutung, Konstruktion und Berechnung) Mykolaiv 1911
Addis, Bill: Building: 3000 years of Design Engineering and Construction. London 2007
Straub, Hans: Die Geschichte der Bauingenieurkunst. Basel 1992
Kurrer, Karl-Eugen: The history of the theory of structures. From arch analysis to computational mechanics. Berlin 2008
Kurrer, Karl-Eugen: Geschichte der Baustatik. Berlin 2002

[13] Straub, Hans: Die Geschichte der Bauingenieurkunst. Basel 1992, p. 206

[14] Kurrer, Karl-Eugen: Geschichte der Baustatik. Berlin 2002, p. 27

[15] Gottgetreu, Rudolf: Lehrbuch der Hochbau-Konstruktionen. Dritter Teil – Eisenkonstruktionen. Berlin 1885

[16] Lorenz, Werner: Die Entwicklung des Dreigelenksystems im 19. Jahrhundert. In: Stahlbau 01/1990, p. 6

[17] cf. 14, p. 27

[18] cf. 14, p. 391

[19] Hinweis von Ines Prokop. Vgl. Prokop, Ines: Eiserne Tragwerke in Berlin. 1850–1925. Einfluss von Berechnungsmethoden und Material auf die Bauwerke. Dissertation at the Universität der Künste. Berlin 2011, pp. 466–468

[20] Scharowsky, Karl: Musterbuch für Eisen-Constructionen. Leipzig 1888

[21] cf. 11

[22] Šuchov, Vladimir G.: Izbrannye trudy. Stroitel'naja mechanika. Pod. red. A. Ju. Išlinskogo. Moscow 1977, pp. 159–169. (Complete German translation by Ottmar Pertschi: Berechnung eines Leuchtturms mit bis zu 68 m lichter Höhe nach dem System des Ingenieurs V. G. Šuchov) Translation Service of the University Library Stuttgart. Ü/570.

[23] Archiv RAN (Archiv der Russischen Akademie der Wissenschaften, Moskau). Op. 1508-83.

[24] cf. 11

[25] Belyi, J., Charičkov, I.: Pamjatniki nauki i techniki. Moscow 1981 (Complete German translation by Ottmar Pertschi: Ein Šuchov Turm in der Stadt Nikolaev – ein Denkmal der Ingenieurskunst). Translation Service of the University Library Stuttgart. Ü/311.

[26] Nowak, Bernd: Die historische Entwicklung des Knickstabproblems und dessen Behandlung in den Stahlbaunormen. Heft 35. Publication of the Instituts für Statik und Stahlbau der Technischen Universität Darmstadt, 1981

[27] Petrov, Dmitrij V.: Železnye vodonapornye bašni. Ich naznačenie, konstrukcii i rasčety. Mykolaiv 1911 (Complete German translation by Ottmar Pertschi: Wasserturm aus Eisen der städtischen Wasserversorgung von Nikolaev mit einem Behälter von 50 Tausend Eimer Fassungsvermögen). Translation Service of the University Library Stuttgart. Ü/626, p. 7

[28] Petrov, Dmitrij V.: Železnye vodonapornye bašni. Ich naznačenie, konstrukcii i rasčety. Mykolaiv 1911 (Complete German translation by Ottmar Pertschi: Wasserausgleichsturm aus Eisen mit einem Fassungsvermögen von 33350 Eimer und wasserundurchlässigem Grundlauf). Translation Service of the University Library Stuttgart. Ü/625, p. 9

[29] Petropavlovskaja, Irina A.: Hyperbolische Gittertürme. In: Graefe, Rainer; Gappoev, Murat; Pertschi, Ottmar: Vladimir G. Šuchov. 1853–1939. Die Kunst der sparsamen Konstruktion. Stuttgart 1990, p. 82

[30] cf. 23

[31] cf. 22, p. 2

[32] cf. 22, p. 3

[33] ibid.p. 3

[34] ibid. p. 5

[35] ibid. p. 6

[36] Šuchov, Vladimir: Rasčet bašni vysotoju 128 mtr (Transliteration into German: Berechnung eines 128 m hohen Turms). Stadtarchiv Nizhny Novgorod

[37] Beckh, Matthias; Barthel, Rainer; Kutnyi, Andrij: Construction and structural behaviour of Vladimir Šuchovs NiGRES tower. In: Sixth International Conference on Structural Analysis of Historic Constructions. Bath 2008, S. 183–190.

Literature

[38] cf. 36, p. 1

[39] Vgl. z. B.: Akademischen Verein Hütte, Abteilung I (ed.): Des Ingenieurs Taschenbuch. Berlin 1902

[40] cf. 25, p. 182

[41] cf. 36, p. 3

[42] Djadjuša, V. A.: Padenie vodonapornoj bašni (Transliteration into German: Der Sturz eines Wasserturms). In: Sanitarnaja technika. Nr. 2-3/1931, pp. 25–30

[43] Popov, G. D. (inž.): Rasčet bašen sistemy Šuchova (Transliteration into German: Berechnung Šuchovscher Wassertürme). In: Stroitel'naja promyšlennost'. Nr. 7/1931, pp. 375–376.

[44] Dinnik, Aleksandr N.: Ustojčivost'uprugich sistem. Glava VII. Različnye slučai ustojčivosti. § 61. Moscow 1950, pp. 128–130 (Complete German translation by Ottmar Pertschi: Stabilität krummliniger Gitter). Translation Service of the University Library Stuttgart. Ü/627

[45] Gorenštejn, B. V.: Rasčet prostranstvennych konstrukcij. Moscow 1959, pp. 146–182 (Complete German translation by Ottmar Pertschi: Berechnung der Gittersysteme V. G. Šuchovs auf Festigkeit, Steifigkeit und Stabilität)

[46] Griškova, N. P.; Lyskov, V. P.; Pen'kov, A. M.: Rasčet bašen sistemy Šuchova na pročnost' i ustojčivost' (Transliteration into German: Festigkeits- und Stabilitätsberechnung der Šuchovschen Konstruktionen). Kharkiv/Dnepropetrovsk 1934

[47] Šuchov, Vladimir: Stropila. Izyskanie racinal'nych tipov prjamolinejnych stropil'nych ferm i teorijaaročnych ferm. Moscow 1897 (Transliteration into German by Ottmar Pertschi: Der Dachverband. Ermittlung der rationellen geradlinigen Dachträger-Typen und Theorie der Bogenbinder). Translation Service of the University Library Stuttgart. Ü/549

[48] Šuchov, Vladimir G.: Stroitel'naja mechanika. Izbrannye trudy. Moscow 1977, pp. 53–64 (Transliteration into German by Ottmar Pertschi: Die Gleichung Eld4y/dx4 = -αy in Aufgaben der Baumechanik, Übersetzungsstelle der Universitätsbibliothek Stuttgart. Ü/336)

[49] Šuchov, Vladimir G.: Stroitel'naja mechanika. Izbrannye trudy. RAN Op. 1508-55 Blatt 1-11. Moscow 1977, pp. 178–185 (Transliteration into German by Ottmar Pertschi: Fabrikgebäude Lys'va. Translation Service of the University Library Stuttgart. Ü/624)

[50] Išlinskij, Aleksandr: Über Šuchovs Beitrag zur Planung und Berechnung von Baukonstruktionen (Transliteration into German by Ottmar Pertschi) In: Šuchov, Vladimir G.: Stroitel'naja mechanika. Izbrannye trudy. Moscow 1977, pp. 4–9 Übersetzungsstelle der Universitätsbibliothek Stuttgart. Ü/247, p. 1

[51] see also: Günther, Daniel: Die hyperbolischen Gitterstabkonstruktionen von V. G. Šuchov. Degree thesis at the Technischen Universität München, 2003 Archiv RAN Op. 1508-82

[52] Petrov, Dmitrij: Železnye vodonapornye bašni. Ich naznačenie, konstrukcii i rasčety. (Transliteration into German: Eiserne Wassertürme. Ihre Bedeutung, Konstruktion und Berechnung). Mykolaiv 1911

NiGRES tower on the Oka

[1] see also:
Beckh, Matthias; Barthel, Rainer; Kutnyi, Andrij: Construction and structural behaviour of Vladimir Šuchovs NiGRES tower. In: Sixth International Conference on Structural Analysis of Historic Constructions. Bath 2008, pp. 183–190 Gappoev, Murat; Graefe, Rainer: Rettungsaktion für Šuchov-Bauten in der Region Nižnji Novgorod. In: Stahlbau 02/2008, p. 99–104

[2] The repair of the tower was initiated and taken forward by Prof. Dr.-Ing. Rainer Graefe and performed on site under the management of Prof. Dr.-Ing. Igor Molev on the Russian side.

[3] Petropavlovskaja, Irina A.: Der Sendeturm für die Radiostation Šabolovka in Moskau. In: Graefe, Rainer; Gappoev, Murat; Pertschi, Ottmar: Vladimir G. Šuchov. 1853–1939. Die Kunst der sparsamen Konstruktion. Stuttgart 1990, p. 100

[4] DIN 1055-4 Einwirkungen auf Tragwerke – Teil 4 Windlasten. 2005-03

[5] DIN 18800-1 Element 749. 2008-11

Résumé

[1] Stachowiak, Herbert: Allgemeine Modelltheorie. Vienna/New York 1973

Towers in comparison

[1] The author wishes to thank Ottmar Pertschi for his help and insight in the analysis of archive materials

[2] Petrov, Dmitrij V.: Železnye vodonapornye bašni. Ich naznačenie, konstrukcii i rasčety (Transliteration into German: Eiserne Wassertürme. Ihre Bedeutung, Konstruktion und Berechnung). Mykolaiv 1911

Archiv RAN (Archiv der Russischen Akademie der Wissenschaften, Moskau). Op. 1508-83

Archiv RAN (Archiv der Russischen Akademie der Wissenschaften, Moskau). Op. 1508-58

Archiv RAN (Archiv der Russischen Akademie der Wissenschaften, Moskau). Op. 1508-82

Archiv RAN (Archiv der Russischen Akademie der Wissenschaften, Moskau). Op. 1508-9

Archiv RAN (Archiv der Russischen Akademie der Wissenschaften, Moskau). Op. 1508-84

Bargmann, Horst: Historische Bautabellen. Düsseldorf 1998

Beckh, Matthias; Musil, Josef; Göttig, Roland: Back to the future – parametric design of hyperboloid lattice towers. In: Göttig, Roland; Schubert, Gerhard (ed.): 3D-Technologien an der Technischen Universität München/Forum 3D. Aachen 2009, pp. 63–74.

Beckh, Matthias: Hyperbolische Turmstrukturen von Vladimir Šuchov. In: Barthel, Rainer (ed.): Denkmalpflege und Instandsetzung. Lectures in winter semester 2008/2009. Chair of Structural Design, Technischen Universität München, pp. 9–23.

Beles, Aurel; Soare, Mircea: Das elliptische und hyperbolische Paraboloid im Bauwesen. Berlin 1970

Beles, Aurel; Soare, Mircea: Berechnung von Schalentragwerken. Wiesbaden/Berlin 1972

Bögle, Annette; Schmal, Peter; Flagge, Ingeborg: leicht weit. Light Structures. Jörg Schlaich, Rudolf Bergermann. Munich 2003

Bögle, Annette: Zur Morphologie komplexer Formen im Bauwesen. Dissertation at the Institute for Lightweight Structures and Conceptual Design, University of Stuttgart, 2005

Chan-Magomedow, Selim O.: Pioniere der sowjetischen Architektur. Dresden 1983

Chudjakov, P.: Novye tipy metalličeskich i derevjannych pokrutij dlja zdanij po sisteme inzenera Šuchova. Tečničeskij sbornik i Vestnik promyšlennosti 1896 (Transliteration into German by Ottmar Pertschi: Neue Metall- und Holzdachtypen für Gebäude nach dem System des Ingenieurs Šuchov. Translation Service of the University Library Stuttgart, Ü/361)

DIN 1055-100 Einwirkungen auf Tragwerke – Teil 100: Grundlagen der Tragwerksplanung, Sicherheitskonzept und Bemessungsregeln. 2001-03

DIN 1055-4 Einwirkungen auf Tragwerke – Teil 4 Windlasten. 2005-03

DIN 18800-1 Stahlbauten. Bemessung und Konstruktion. 2008-11

DIN 18800-2 Stahlbauten. Stabilitätsfälle – Knicken von Stäben und Stabwerken. 2008-11

DIN 18800-4 Stahlbauten. Stabilitätsfälle – Schalenbeulen. 2008-11

DIN EN 1990 Eurocode: Grundlagen der Tragwerksplanung. Deutsche Fassung. 2002-10

Flügge, Wilhelm: Statik und Dynamik der Schalen. Berlin 1981

Foerster, Max: Taschenbuch für Bauingenieure. Berlin 1928

Foerster, Max: Die Eisenkonstruktionen der Ingenieur-Hochbauten. Ergänzungsband zum Handbuch der Ingenieurwissenschaften. Leipzig 1902

Föppl, August: Das Fachwerk im Raume. Leipzig 1892

Föppl, August: Die Mechanik im neunzehnten Jahrhundert, in Bericht über die Königlich Technische Hochschule zu München. Munich 1892

Gössel, Peter; Leuthäuser, Gabriele: Architektur des 20. Jahrhunderts. Cologne 2005

Gottgetreu, Rudolph: Lehrbuch der Hochbau-Konstruktionen. Dritter Teil – Eisenkonstruktionen. Berlin 1885

Graefe, Rainer: Filigrane Hallen – Paxton und Šuchov. In: Schunk, Eberhard (ed.): Beiträge zur Geschichte des Bauingenieurwesens – Hallen. Lehrstuhl für Baukonstruktion, Technischen Universität München 1997

Hampe, Erhard: Rotationssymmetrische Flächentragwerke. Berlin 1981

Harris, John; Stocker, Horst: Handbook of Mathematics and Computational Science. New York 1998

Hartung, Giselher: Eisenkonstruktionen des 19. Jahrhunderts. Munich 1983

Hoff, Robert: Meisterwerke der Ingenieurbaukunst. Köln 1998

Holgate, Alan: The art of structural engineering. The work of Jörg Schlaich and his team. Stuttgart 1997

Kohlmaier, Georg; von Sartory, Barna: Das Glashaus – ein Bautypus des 19. Jahrhunderts. Munich 1981

Königer, Otto: Allgemeine Baukonstruktionslehre – Die Konstruktionen in Eisen. Hannover 1902

Kovel'man, Grigorjewitsch M.: Trudy po istorii techniki. Materialy pervogo soveščanija po istorii techniki. Moscow 1954, pp. 64–88. (Transliteration into

German by Ottmar Pertschi: Vladimir G. Šuchov, der größte russische Ingenieur (1853–1939). Translation Service of the University Library Stuttgart. Ü/208)

Kovel'man, Grigorjewitsch. M.: Tvorčestvo početnogo akademika inženera Vladimira Grigor'eviča Šuchova. Moscow 1961

Kurrer, Karl-Eugen: Grace and law: The spatial framework from Föppl to Mengeringhausen. In: Essays in the history of the theory of structures – In honour of Jacques Heyman. Madrid 2005

Le Corbusier: Ausblick auf eine Architektur. Brunswick/Wiesbaden 1982

Lehmann, Christine; Maurer, Bertram: Karl Culmann und die graphische Statik. Berlin 2006

Mel'nikov, N. P.: V. G. Šuchov – osnovopoložnik otečestvennoj konstruktorskoj školy. In: V. G. Šuchov – vydajuščijsja inžener i učenyj. Trudy Ob'edinennoj naučnoj sessii Akademii nauk SSSR. Posvjaščennoj naučnomu i inženernomu tvorčestvu počcetnogoi akademika V. G. Šuchova. Moscow 1984

Mišin, V. P.: Metalličeskie konstrukcii akademika V. G. Šuchova. Sostavitel' I. A. Petropavlovskaja. Moscow 1990

Meyer, Alfred G.: Eisenbauten – Ihre Geschichte und Ästhetik. Berlin 1907

Nerdinger, Winfried: Konstruktion und Raum in der Architektur des 20. Jahrhunderts. Munich 2002

Pare, Richard: The lost Vanguard – Russian Modernist Architecture 1922–1932. New York 2007

Petersen, Christian: Dynamik der Baukonstruktionen. Brunswick/Wiesbaden 2000

Picon, Antoine: L'Art de l'ingénieur – constructeur, entrepreneur, inventeur. Paris 1997

Ricken, Herbert: Der Bauingenieur. Geschichte eines Berufes. Berlin 1994

Ricken, Herbert: Otto Intze, in Zur Geschichte der Bauingenieurkunst und-wissenschaft. Berlin 1998

Schädlich, Christian: Das Eisen in der Architektur des 19. Jahrhunderts. Beitrag zur Geschichte eines neuen Baustoffs. Postdoctoral thesis. Universität Weimar, 1967 (not published)

Schlaich, Jörg; Heinle, Erwin: Kuppeln aller Zeiten, aller Kulturen. Stuttgart 1996

Schober, Hans: Die Masche mit der Glaskuppel. In: db 10/1994, pp. 152–163.

Schwedler, Johann Wilhelm: Die Construction der Kuppeldächer. Zeitschrift für Bauwesen. 16. Jahrgang 1866. pp. 7–34

Schwedler, Johann Wilhelm: Eiserne Dachkonstruktion über Retortenhäuser der Gas-Anstalten zu Berlin. 19. Jahrgang 1869, pp. 66–70

Stachowiak, Herbert: Allgemeine Modelltheorie. Vienna/New York 1973

Sören, Stephan; Jaime, Sánchez-Alvarez; Klaus, Knebel: Stabwerke auf Freiformflächen. In: Stahlbau 08/2004, pp. 562–572

Torroja, Eduardo: Logik der Form. Munich 1961

Wriggers, Peter: Nichtlineare Finite-Elemente Methoden. Berlin/Heidelberg/New York 2001

Picture credits

Photographs not specifically credited are taken from the archives of the magazine "DETAIL, Review of Architecture". Despite intensive endeavours, we were unable to establish copyright ownership in just a few cases; howevercopyright is assured. Please notify us accordingly in such instances.
All drawings not listed here were prepared by the author.

Cover
Matthias Beckh, Munich

Foreword
Fig. 1   RAN archive (Archive of the Russian Academy of Sciences in Moscow) Op. 1508-84

Introduction
Fig. 2   Matthias Beckh, Munich

Building with hyperbolic lattice structures
Fig. 1   Crystal Palace Exhibition, Illustrated Catalogue, London 1851
Fig. 2   Renaissance der Bahnhöfe, Wiesbaden 1996, p. 266
Fig. 3   Matthias Beckh, Munich
Fig. 4   Graefe, Rainer; Gappoev, Murat; Pertschi, Ottmar: Vladimir G. Šuchov. 1853–1939. Die Kunst der sparsamen Konstruktion. Stuttgart 1990, p. 46
Fig. 5   Chair of Structural Design, Technische Universität München
Fig. 6   Heide Wessely, Munich
Fig. 7   see Fig. 4, p. 32
Fig. 8   see Fig. 4, p. 177
Fig. 9   Nachlass Reinhard Mannesmann, Archiv des Deutschen Museums
Fig. 10a Vinogradowa, T.; Petrow, I.: Nižnij Novgorod Photography. Nizhny Novgorod 2007
Fig. 10b Vladimir F. Shukhov, Moscow
Fig. 11 a see Fig. 4, p. 80
Fig. 11 b see Fig. 4, p. 80
Fig. 11 c see Fig. 4, p. 81
Fig. 11 d see Fig. 4, p 80
Fig. 12   see Fig. 4, p. 95
Fig. 13   Torroja, Eduardo: Logik der Form. Munich 1961, p. 250
Fig. 14   Jodidio, Philip (ed.): I. M. Pei Compete Works. New York 2008, p. 43
Fig. 15   see Fig. 3
Fig. 16   Information Based Architecture, Amsterdam
Fig. 17   see Fig. 3

Structural analysis and calculation methods
Fig. 12   Vladimir F. Shukhov, Moscow
Fig. 13   nach Graf, Jürgen: Entwurf und Konstruktion von Translationsnetzschalen. Dissertation at the Institute of Structural Design. University of Stuttgart, 2002, p. 126
Fig. 14   nach Ramm, Ekkehard: Geometrisch nichtlineare Elastostatik und Finite Elemente. Bericht Nr. 76-2. Institute for Structural Mechanics, University of Stuttgart, 1976, p. 78
Fig. 15   nach Werkle, Horst: Finite Elemente in der Baustatik. Statik und Dynamik der Stab- und Flächentragwerke. Wiesbaden 2008, p. 42
         und Ramm, Ekkehard: Geometrisch nichtlineare Elastostatik und Finite Elemente. Bericht Nr. 76-2. Institute for Structural Mechanics, University of Stuttgart, 1976, p. 71
Fig. 16   nach Petersen, Christian: Dynamik der Baukonstruktionen. Brunswick/Wiesbaden 2000, p. 38
Fig. 17   see Fig. 13, p. 125
Fig. 18   nach Gioncu, Victor; Balut, Nicolae: Instability behavior of single layer reticulated shells. In: International Journal of Space Structures. Vol 7. No. 4/1992, pp. 243–252
Fig. 21   Ansys Help System, 2007
Fig. 26   Matthias Beckh, Munich

Relationships between form and structural behaviour
Fig. 24   Hoheisel, Meike: Einfluss der Formparameter auf das Tragverhalten hyperbolischer Stabwerke. Master Theses on the subject of civil engineering. Chair of Structural Design/Chair of Structural Analysis, Technische Universität München, 2010

Design and analysis of Shukhov's towers

Fig. 1   Matthias Beckh, Munich
Fig. 2a   Werth, Jan: Ursachen und technische Voraussetzungen für die
          Entwicklung der Wasserhochbehälter. In: Becher, Bernd und Hilla:
          Industriearchitektur des neunzehnten Jahrhunderts. Die Architektur
          der Förder- und Wassertürme. Munich 1971, p. 337
Fig. 2b   see Fig. 2a, p. 346
Fig. 2c   see Fig. 2a, p. 349
Fig. 3   see Fig. 2a, p. 350
Fig. 4   see Fig. 2a, p. 351
Fig. 5   see Fig. 2a, p. 355
Fig. 6a   see Fig. 2a, p. 356
Fig. 6b   see Fig. 2a, p. 358
Fig. 7   see Fig. 2a, p. 367
Fig. 8   see Fig. 2a, p. 364
Fig. 9   Merkl, Gerhard et al.: Historische Wassertürme – Beiträge zur Tech-
          nikgeschichte von Wasserspeicherung und Wasserversorgung.
          Munich/Vienna 1985, p. 120
Fig. 10   Maggi-Werke, Singen
Fig. 11   Petrov, Dm. V.: Železnye vodonapornye bašni. Ich naznačenie, konst-
          rukcii i rasčety (Transliteration into German: Eiserne Wassertürme.
          Ihre Bedeutung, Konstruktion und Berechnung). Mykolaiv 1911, p. 95ff.
Fig. 13   Šuchov, V. G.: Izbrannye trudy. Stroitel'naja mechanika. Pod. red. A.
          Ju. Išlinskogo. Moscow 1977, pp. 159–169. (Complete German
          translation by Ottmar Pertschi: Berechnung eines Leuchtturms mit bis
          zu 68 m lichter Höhe nach dem System des Ingenieurs V. G. Šuchov.
          Translation Service of the University Library Stuttgart. Ü/570, p. 4)
Fig. 14   Petrov, Dmitrij V.: Železnye vodonapornye bašni. Ich naznačenie,
          konstrukcii i rasčety. Mykolaiv 1911. (Complete German translation
          by Ottmar Pertschi: Wasserturm aus Eisen der städtischen Wasser-
          versorgung von Nikolaev mit einem Behälter von 50 Tausend Eimer
          Fassungsvermögen. Translation Service of the University Library
          Stuttgart. Ü/626, p. 8)
Fig. 15   Shuhkov archive of Rainer Graefe, Innsbruck.
Fig. 16   see Fig. 11, p. 72
Fig. 17   Oleksandr Serdyuk, Cherson
Fig. 18   Oleksandr Serdyuk, Cherson
Fig. 19   see Fig. 13, p. 10
Fig. 20   RAN archive (Archive of the Russian Academy of Sciences in Moscow).
          Op. 1508-83
Fig. 21   see Fig. 20
Fig. 23   see Fig. 20
Fig. 24   see Fig. 13, p. 6
Fig. 25   see Fig. 20
Fig. 26a   RAN archive (Archive of the Russian Academy of Sciences in Moscow).
          Op. 1508-58
Fig. 26b   Shuhkov archive of Rainer Graefe, Innsbruck
Fig. 26c   see Fig. 26b
Fig. 27   Šuchov, Vladimir G.: Rasčet bašni vysotoju 128 mtr (Transliteration into
          German: Berechnung eines 128 m hohen Turms). Stadtarchiv Nizhny
          Novgorod
Fig. 28a   Djadjuša, V.A.: Padenie vodonapornoj bašni (Transliteration into German:
          Der Sturz eines Wasserturms). In: Sanitarnaja technika. Nr. 2-3/1931, pp.
          25–30
Fig. 28b   Griškova, N.P.; Lyskov, V.P.; Pen'kov, A.M.: Rasčet bašen sistemy
          Šuchova na pročnost' i ustojčivost' (Transliteration into German: Festig-
          keits- und Stabilitätsberechnung der Šuchovschen Wassertürme). Khar-
          kiv/Dnepropetrovsk 1934
Fig. 29   see Fig. 1
Fig. 30   Popov, G.D. (inž.): Rasčet bašen sistemy Šuchova (Transliteration into
          German: Berechnung Šuchovscher Wassertürme). In: Stroitel'naja
          promyšlennost'. Nr. 7/1931, pp. 375–376.
Fig. 31   Gorenštejn, B.V.: Rasčet prostranstvennych konstrukcij. Moscow 1959,
          pp. 146–182 (Complete German translation by Ottmar Pertschi: Berech-
          nung der Gittersysteme V.G. Šuchovs auf Festigkeit, Steifigkeit und Sta-
          bilität)
Fig. 32   RAN archive (Archive of the Russian Academy of Sciences in Moscow).
          Op. 1508-82
Fig. 34   Shuhkov archive of Rainer Graefe, Innsbruck

Fig. 35   see Fig. 34
Fig. 36   see Fig. 34
Fig. 37   see Fig. 34
Fig. 38   Chair of Structural Design, Technische Universität München

NiGRES tower on the Oka

Fig. 1   Lopatto, A.E.: Vladimir Grigor'evich Šuchov – vydajuščijsja russkij
          inžener. Moscow 1951
Fig. 2a   Matthias Beckh, Munich
Fig. 2b   Igor Molev, Nizhny Novgorod
Fig. 3   Gerhard Weiss, Munich
Fig. 4   Shuhkov archive of Rainer Graefe, Innsbruck
Fig. 9   Chair of Structural Design, Technische Universität München
Fig. 10   see Fig. 2a
Fig. 12   see Fig. 2a
Fig. 20   see Fig. 2a
Fig. 24   Šuchov, Vladimir G.: Rasčet bašni vysotoju 128 mtr (Transliteration into
          German: Berechnung eines 128 m hohen Turms). Stadtarchiv Nizhny
          Novgorod

Résumé

Fig. 1   Matthias Beckh, Munich

Towers in comparison

p. 115 oben links   Vinogradova, T.; Petrow, I.: Nižnij Novgorod Photography.Niz-
                    hny Novgorod 2007
p. 115 oben rechts   Shuhkov archive of Rainer Graefe, Innsbruck
p. 115 unten   Shuhkov archive of Rainer Graefe, Innsbruck
p. 116 oben links   Graefe, Rainer; Gappoev, Murat; Pertschi, Ottmar: Vladimir G.
                    Šuchov. 1853–1939. Die Kunst der sparsamen Konstruktion.
                    Stuttgart 1990, p. 80
p. 116 oben rechts   Shuhkov archive of Rainer Graefe, Innsbruck
p. 116 unten   Shuhkov archive of Rainer Graefe, Innsbruck
p. 117 oben links   Petrov, Dmitrij V.: Železnye vodonapornye bašni. Ich
                    naznačenie, konstrukcii i rasčety. (Transliteration into German:
                    Eiserne Wassertürme. Ihre Bedeutung, Konstruktion und
                    Berechnung). Mykolaiv 1911, p. 72
p. 117 oben rechts   Graefe, Rainer et al.: Vladimir G. Šuchov. 1853–1939, p. 80
p. 117 unten links   Graefe, Rainer et al.: Vladimir G. Šuchov. 1853–1939, p. 80
p. 117 unten rechts Graefe, Rainer et al.: Vladimir G. Šuchov. 1853–1939, p. 81
p. 118 oben   Graefe, Rainer et al.: Vladimir G. Šuchov. 1853–1939, p. 81
p. 118 unten   Shuhkov archive of Rainer Graefe, Innsbruck
p. 119 oben   Graefe, Rainer et al.: Vladimir G. Šuchov. 1853–1939, p. 80
p. 119 unten   Djadjuša, V.A.: Padenie vodonapornoj bašni (Transliteration into Ger-
                    man: Der Sturz eines Wasserturms). In: Sanitarnaja technika. Nr.
                    2-3/1931, p. 28

Images introducing chapters (Matthias Beckh)

p. 10   Buckling shape of a hyperbolic lattice tower in plan
p. 92   NiGRES tower on the Oka, view up inside the tower, Dzerzhinsk (RUS)
          1929
p. 138   NiGRES tower on the Oka, view up the access ladder, Dzerzhinsk

About the author

Matthias Beckh (born 1974)
Studied civil engineering at Technical University (TU) Munich
Development of measures to safeguard objects of cultural heritage in earthquake
zones in underdeveloped regions (as part of the Postgraduate Program in International Affairs supported by the German National Academic Foundation and the Robert Bosch Stiftung)
1999–2005 worked at Robert Silman Associates, New York and at Guy Nordenson
and Associates, New York
2005–2013 Research Fellow at the Institute of Structural Design at the Faculty for
Architecture, TU Munich and own projects in the field of new and old buildings/refur-
bishment (including in Nepal and Bhutan)
Since 2013 works for Buro Happold, Munich/Berlin

Notation

■	Compression
■	Tension

R_U	Bottom radius of the hyperbolic lattice structure
R_O	Top radius of the hyperbolic lattice structure
K_F	Shape parameter; ratio RU/RO
H	Height
IR	Number of intermediate rings
n	Number of vertical member pairs
n_{SP}	Number of vertical member intersection points
ψ	Central angle
φ	Rotation angle of the vertical members in plan
β	Angle between start of vertical member and waist in plan
γ	Angle between waist and end of vertical member in plan
α	Angle in the triangle in plan
R_T	Waist radius
H_T	Waist height
L	True length of vertical member
G	Length of vertical member in plan
ε	Inclination of the member relative to the vertical axis

F_S	Member force
F_H	Horizontal component of the member force
$F_{N,X}$	Force component normal to the ring
$F_{T,X}$	Force component tangential to the ring
δ	Angle between the line connecting the bottom support points of a pair of vertical members and the directions of the vertical members in plan

K	Stiffness matrix
u	Vector of the node displacements
F	Vector of the external loads
P	External load
P_{crit}	Ultimate load
P_{k1}	Ultimate load with perfect geometry, linear or non-linear calculation
P_{k2}	Ultimate load with imperfect geometry, linear or non-linear calculation
p_G	Resulting wind pressure on the lattice per linear unit of length (original calculations)
$F_{W,Int}$	Resulting wind load on the Intze tank (original calculations)
Q	Transverse force or maximum vertical member force (original calculations)
φ	Reduction of the permissible stress resulting from a buckling check using the Laissle-Schübler method (original calculations)
W_{Ring}	Resulting wind load on the main and intermediate rings (original calculations)

Bucket	Russian measure of volume, equates to 12.3 litres
Pood	Russian measure of weight, equates 16.38 kg
Foot	Measure of length, equates to 0.3048 m

Index

A
Adziogol lighthouse   75ff.
Alexander II, Tzar   17
All-Russia Exhibition Nizhny Novgorod   8, 17f., 21
Ansys   45f., 60
Arched girder   17

B
Bari, Alexander V.   17
Belánger, François-Joseph   14
Bell, Alexander Graham   15
Bending, inextensional   32f.
Brunet, François   14
Buckling   40

C
Calculation, linear   40f.
Calculation, non-linear   42
Circular cylindrical shell   50f.
Conic sections   24
Contamin, Victor   15
Cross section, tubular   37, 73, 77f., 86, 112
Crystal Palace   14
Culmann, Carl   15, 70

D
Design process   89ff., 112
Dinnik, A. N.   85
Dnipropetrovsk (water tower)   60ff., 84f., 135
Dutert, Charles Louis Ferdinand   15

E
Eiffel, Gustave   15
Eigenvalue analysis   40f.

F
Failure, stability   38ff.
Failure, stress   38f.
Finite-element analysis   45f., 103, 110
Flat-bottomed tanks   66, 90
Foster, Norman   18

G
Gappoev, Murat   8, 10, 17
Gaussian curvature   28
Gorenšteijn, B. V.   86
Graefe, Rainer   8, 10, 17
Gridshells   15, 17f., 32

H
Horizontal load transfer   34ff.

Hyperboloid of rotation   27, 77
Hyperboloid, one-sheeted   24, 45, 50

I
Imperfection shape   43ff., 49
Individual member buckling   40
Inextensional bending   32
Intermediate rings   53, 54ff., 78, 87
Intze, Otto   66ff.
Intze tank   66ff., 136ff.

J
Yaroslavl (two-storey water tower)   22, 79f., 137

K
Kovel'man, Grigorij   10, 19

L
Linear calculation   40f.
Load assumptions   47
Load capacity calculation   38ff., 49
Load-displacement curve   40ff., 49, 61ff.
Load transfer, horizontal   34ff., 45, 103ff.
Load transfer, vertical   32ff., 103

M
Maillart, Robert   15
Mannesmann, Max and Reinhard   21
Member cross section   46
Member forces   32ff., 72f.
Moscow   17, 22
Multilayer construction   46f., 56f.

N
Navier, Claude Louis   15, 70
Nervi, Pier Luigi   15
Newton-Raphson method   42
NiGRES tower on the Oka   23, 79ff., 96ff.
Mykolaiv (water tower)   60, 62, 70ff., 126, 137
Nizhny Novgorod (water tower)   22, 39, 60f., 122, 136
Node load   37
Node snap-through   40
Node stiffness   47, 58
Non-linear calculation   42
Non-linear stability analysis   41ff., 49, 52f., 60ff.

O
Original calculations   70ff., 84ff., 110, 112

P
Parametric studies   45ff., 50ff.
Patent application No. 1896   19f.

Pei, Ieoh Ming   23
Pertschi, Ottmar   10, 17
Popov, G.D.   85

R
Radius size   46, 53f.
Rice, Peter   17
Rotation angle   29f., 50, 53

S
Shabolovka radio tower   8f., 22f., 96, 110
Schlaich, Jörg   17
Schwedler, Johann Wilhelm   15, 70
Second order surface   24
Shape parameter   29, 50ff.
Sickle girder   14
Stability analysis   40, 41ff.
Stability failure   38ff.
Stress-controlled failure   40
Stress failure   38f.
Structural behaviour   32ff., 105ff.
Shukhov, Vladimir G.   15ff., 112f.
Suspended roof   18f.
Suspended tank   66
Suspended-span girders   15

T
Telescopic method   96ff.
Three-pinned arches   15
Tyumen (water tower)   60, 63, 75, 127
Top load   34f.
Torroja, Eduardo   23
Torsion   38
Tubular cross section   37, 73, 77f., 85f., 112
Turner, Richard   14

U
Ultimate load capacity   38ff., 49

V
Van de Velde, Henry   15
Vertical load transfer   32ff., 103
Vertical member arrangement   28ff., 46f., 56, 107
Vyksa   17

W
Water towers   22, 60ff., 122ff.
Wind loads   47f., 77ff., 85f., 103ff., 110, 113

Acknowledgement

This publication is based on my PhD thesis, which I received from the Technische Universität München (TUM). For the supervision of my thesis I would like to expressly thank Rainer Barthel and Rainer Graefe. Through their support I was able to pursue this field of research during my time at the Department of Structural Design and the Faculty of Architecture at the TUM. Ottmar Pertschi provided me with generous support in the analysis of archive materials.

Furthermore, I would like to sincerely thank the following people for their support, suggestions and encouragement:
Kai-Uwe Bletzinger, Jürg Conzett, Matthias Eckert, Sergej Fedorov, Meike Hoheisel, Karl-Eugen Kurrer, Werner Lorenz, Igor Molev, Ekaterina Nozhova, Ines Prokop, and all the colleagues from the research project: "Konstruktionswissen der frühen Moderne – Suchovs Strategien des sparsamen Eisenbaus".

Moreover, I would like to thank Cornelia Hellstern and Eva Schönbrunner from Detail for their care and dedication in the preparation of this book.

February 2014
Matthias Beckh

Printed and bound by CPI Group (UK) Ltd, Croydon, CR0 4YY

28/10/2024

14581379-0001